D1594562

*Now available in a lower priced paperback edition in the Wiley Classics Library.

INTRODUCTION TO
MODERN SET THEORY

INTRODUCTION TO MODERN SET THEORY

JUDITH ROITMAN
Department of Mathematics
University of Kansas
Lawrence, Kansas

WILEY

A Wiley-Interscience Publication

JOHN WILEY & SONS

New York • Chichester • Brisbane • Toronto • Singapore

QA
248
R645
1990

Copyright © 1990 by John Wiley & Sons, Inc.

All rights reserved. Published simultaneously in Canada.

Reproduction or translation of any part of this work
beyond that permitted by Section 107 or 108 of the
1976 United States Copyright Act without the permission
of the copyright owner is unlawful. Requests for
permission or further information should be addressed to
the Permissions Department, John Wiley & Sons, Inc.

Library of Congress Cataloging in Publication Data:

Roitman, Judith.
  Introduction to modern set theory/Judith Roitman.
  p.   cm.—(Pure and applied mathematics) (John Wiley & Sons)
  "A Wiley-Interscience publication."
  Bibliography: p.
  Includes index.
  ISBN 0-471-63519-7
  1. Set theory. 1. Title.
QA248.R645 1989   89-33698
511.3'22—dc20   CIP

Printed in the United States of America

10 9 8 7 6 5 4 3 2

3 3001 00755 7142

To my father,
who loves mathematics

PREFACE

When, in early adolescence, I first saw the proof that the real numbers were uncountable, it seemed the most wonderful thing in the world to me, and I found it quite strange that the rest of the world did not share my enthusiasm. Later, finding out that set theorists could actually prove some basic mathematical questions to be unanswerable and that large infinite numbers could effect the structure of the reals, I was even more amazed that the world did not beat a path to the set theorists' door.

More years later than I care to admit, this book is my response. I wrote it in the firm belief that set theory is good not just for set theorists, but for many mathematicians, and that the earlier a student sees the particular point of view that we call modern set theory, the better.

It is designed for a one-semester course in set theory at the advanced undergraduate or beginning graduate level. It assumes no knowledge of logic, and no knowledge of set theory beyond the vague familiarity with curly brackets, union, and intersection usually expected of an advanced mathematics student. It grew out of my experience teaching this material in a first-year graduate course at the University of Kansas over many years. It is aimed at two audiences—students who are interested in studying set theory for its own sake and students in other areas who may be curious about applications of set theory to their field. While a one-semester course with no logic as a prerequisite cannot begin to tell either group of students all they need to know, it can hope to lay the foundations for further study. In particular, I am concerned with developing the intuitions that lie behind modern, as well as classical, set theory, and with connecting set theory with the rest of mathematics.

Thus, three features are the full integration into the text of the study of models of set theory, the use of illustrative examples both in the text and in the exercises, and the integration of consistency results and large cardinals into the text when appropriate (for example, when cardinal exponentiation is introduced). An attempt is made to give some sense of the history of the subject, both as motivation, and because it is interesting in its own right.

The first chapter is an introduction to partial orders and to well- ordered sets, with a nod to induction on $\mathbb{N}$ and a word about models of first- order theories.

The second chapter introduces the basic set operations and all the axioms except for regularity and choice. Models for each axiom are defined on an ad hoc basis as the axiom is introduced, and many of the exercises involve properties of these models. This serves not only to orient the student towards thinking in terms of models, but also provides reasonably sophisticated material to help the student look carefully at basic concepts.

The third chapter is on regularity and choice. It includes a section on transitive sets and a brief introduction to the ordinals. Examples of how to prove theorems using various versions of the axiom of choice are given.

The fourth chapter is a glimpse of how to code mathematics into the language of set theory.

The fifth chapter is on ordinals and cardinals, including ordinal and cardinal arithmetic.

The sixth chapter is on models: The proof that a strongly inaccessible cardinal gives rise to a model of set theory, and a brief discussion of L.

The seventh chapter is on infinite combinatorics. Topics discussed include partition calculus, trees, measurable cardinals, Martin's axiom, stationary sets, and $\diamondsuit$. Its sections are not entirely independent; for example, weakly compact cardinals are introduced in the first section, explored further in the second, and appear again in the third. The length of discussion of the topics in this chapter is not determined by importance (very little is said about measurable cardinals, for instance, and stationary sets do not appear until the very end) but by pedagogical considerations. Thus, more difficult topics are given shorter shrift, and emphasis is placed on connecting small cardinals with large and exploring fairly simple situations which give rise to consistency results.

If this book is used as a text, the core chapters are the first, second, third, and fifth. Although I am fond of the order and content of the other chapters, someone else teaching out of this book may easily disagree. The level of difficulty of various sections varies enormously, and some students may not be able to handle the more difficult material in chapters 6 and 7. I would expect that most of the audience it is aimed at would not be able to finish it in one semester—chapter 7 is to be regarded as a smorgasbord, rather than a sit-down dinner.

Finally, I hope that a reader completing this text will not only know a good deal of basic set theory, but will have a sense of the richness of the field and of both the pervasiveness and unavoidable nature of mathematical uncertainty.

Many mathematicians have added greatly to this work, and I would like to thank them for it. Much of what is good in this book is due to their help, while all of the errors, whether of omission or commission, are mine. Doug Cenzer, Jean Larson, Jack Porter, and Bill Fleissner used early drafts in the classroom; their comments were crucial in shaping later ones. In addition, the following mathematicians made extensive and valuable suggestions on various versions of the manuscript: Jim Baumgartner, Jim Henle, Istvan Juhasz, Aki Kanamori, Arnold Miller, Peter Nyikos, and Chaz Schwindlein. Many graduate students suffered through various incarnations of this book with me; in particular Tim LaBerge and Steve Schwalm did yeoman service. The technical typist, Sharon Gumm, deserves thanks for her speed and patience, a rare combination in humans. The Wiley production staff deserves special thanks. Finally, I would like to thank my husband, Stanley Lombardo, for his support and love.

JUDITH ROITMAN

CONTENTS

1

SOME MATHEMATICAL PRELIMINARIES

INTRODUCTION

The reader is probably used to picking up a mathematical textbook and seeing the first chapter entitled something like "Set-theoretical prerequisites." Such a chapter usually contains a quick review or an overview of the relevant set theory, from something as simple as the definition of the union of two sets to something as complicated as the definitions of countable and uncountable sets. Since this is a set-theory text, we reverse the usual procedure by putting in the first chapter some mathematics that will prove essential in the serious study of set theory: partially ordered and linearly ordered sets, equivalence relations, well-ordered sets, and induction and recursion. We also point (hopefully not too vaguely) in the direction of mathematical logic with some discussion of first-order theories and models.

Why these topics in particular?

The spine of the set-theoretic universe, and the most essential class of objects in the study of set theory, is the class of ordinals. One of the basic properties of an ordinal is that it is a well-ordered set. An acquaintance with various examples and properties of well-ordered sets is essential to the study of ordinals.

Two of the basic techniques of set theory are transfinite induction and transfinite recursion, which are based on induction and recursion on the natural numbers.

When set theory is applied to the rest of mathematics, the methodology often used is to reduce the original question to a question in the area known as infinite combinatorics. The theories of partially ordered sets and of equivalence relations are essential elements of combinatorics.

When a mathematical question can be settled only by set-theoretic techniques, that is often because consistency results are involved. This revolutionary method of set theory can be understood only with some reference to mathematical logic, in particular to models of first-order theories. An understanding of models is also crucial to the study of constructibility and large cardinals.

Thus the choice of topics in this chapter.

A brief word about the approach. Chapter 1 is written in ordinary mathematical style without set-theoretical formality (compare the definition of partial order in section 1.1 with the formal definition in section 2.3). The reader is assumed to be familiar with set-theoretic notation as found in most advanced mathematical texts, and we will make use of it throughout. The reader is also assumed to be familiar with the standard body of basic mathematics, e.g., the basic properties of the natural numbers, the integers, the rationals, and the reals.

SECTION 1.1. PARTIALLY ORDERED SETS

The most essential combinatorial idea we need is that of a partially ordered set. Here are the axioms defining a partial order $\leq$ on a set X:

For all $x, y, z \in X$,
P1 (Reflexive). $x \leq x$
P2 (Antisymmetric). If $x \leq y$ and $y \leq x$ then $x = y$
P3 (Transitive). If $x \leq y$ and $y \leq z$, then $x \leq z$.

As shorthand, we say $x < y$ (x is strictly less than y) if $x \leq y$ and $x \neq y$. If $\leq$ partially orders X, we call X a partially ordered set under $\leq$ and say $<$ strictly orders X. As will become clear from the examples, a set can have many different partial orders imposed upon it. There are two ways of referring to a set X with a partial order $\leq$. The first is to talk about X under $\leq$. But when it is clear from the context which partial order on X we are talking about, we just refer to the partially ordered set X. Let us check that certain structures are in fact partial orders.

Example 1. The set of positive natural numbers[†] $\mathbb{N}^+ = \{1, 2, 3, \ldots\}$, where we define $n \leq_D k$ iff n divides k. Is $\leq_D$ a partial order on $\mathbb{N}^+$?

[†]$\mathbb{N}$ is the set of natural numbers, $\{0, 1, 2, \ldots\}$.

Check for P1: Every n divides n, so each $n \leq_D n$.

Check for P2: If n divides k then $n \leq k$ (where $\leq$ is the usual order). If k divides n then $k \leq n$. We know that $n \leq k$ and $k \leq n$ implies $n = k$. Hence $k \leq_D n$ implies $n = k$.

Check for P3: If $n \leq_D k$ then $k = in$ for some i. If $k \leq_D m$, then $m = jk$ for some j. So if $n \leq_D k$ and $k \leq_D m$, there are i, j with $m = jk = jin$. Hence $n \leq_D m$.

Notice that the i, j, whose existence was needed for the proof of P3, must come from $\mathbb{N}^+$. The fact that $2 = \frac{2}{3}(3)$ does not imply $2 \leq_D 3$.

Example 2. Let X be any set and define $x \leq_E y$ iff $x = y$ for all $x, y \in X$. Is $\leq_E$ a partial order on X?

Check for P1: Since each $x = x$, each $x \leq_E x$.

Check for P2: If $x \leq_E y$, then $x = y$. So if $x \leq_E y$ and $y \leq_E x$, then $x = y$.

Check for P3: If $x \leq_E y$ and $y \leq_E z$, then $x = y$ and $y = z$, so $x = z$, so $x \leq_E z$. Thus P3 holds.

Example 2 is instructive. Partial orders may in fact have very little structure.

Example 3. Let X be any collection of sets and for all $x, y \in X$ define $x \leq_S y$ iff $x \subset y$. (Recall from your previous studies that $x \subset y$ iff, for all z, if $z \in x$ then $z \in y$. In particular, $x \subset x$. You may have seen $\subset$ written as $\subseteq$. Either convention is acceptable once it has been agreed upon. In accordance with most set-theoretic usage we adopt the former.)

The proof that example 3 is a partial order is left to the reader.

Example 4. Consider the set of real numbers $\mathbb{R}$. The usual order on $\mathbb{R}$ can be defined directly from the algebraic structure of $\mathbb{R}$. First define x to be nonnegative iff there is some $y \in \mathbb{R}$ with $y^2 = x$ (i.e., "nonnegative" means "has a real square root"). Then define $x \leq y$ iff $y - x$ is nonnegative. Finally show that $\leq$ is a partial order. The reader is challenged to prove this algebraically. (Aside from the usual field axioms, you should assume that every sum of squares has a square root and if either a or b are nonzero then so is $a^2 + b^2$.)

Example 5. Let X be all finite sequences of 0's and 1's. We define, for σ,

$\tau \in X$, $\sigma \le \tau$ iff τ extends σ or $\tau = \sigma$. For example, 01001 extends 0100, which extends 010, which extends 01, which extends 0. So $0 \le 01 \le 010 \le 0100 \le 01001$. Notice that if $\sigma \le \tau$ then the length of σ (abbreviated $l(\sigma)$) is less than or equal to the length of τ, $l(\tau)$; i.e., if $\sigma \le \tau$ then $l(\sigma) \le l(\tau)$. Also if we define $\widehat{\sigma \tau}$ to mean σ followed by τ (e.g., $\widehat{010\,10} = 01010$), then if $\sigma \le \tau$ and $\sigma \ne \tau$ there is some ρ with $\widehat{\sigma \rho} = \tau$. We say "$\sigma$ concatenated with τ" for $\widehat{\sigma \tau}$. With this in mind, we show that this is a partial order:

     Check for P1:   Immediate from the definition.

     Check for P2:   If $\sigma \le \tau$, then $l(\sigma) \le l(\tau)$. If $\tau \le \sigma$, then $l(\tau) \le l(\sigma)$. Thus $\sigma \le \tau \le \sigma$ implies $l(\sigma) = l(\tau)$. Hence τ does not extend σ; hence $\tau = \sigma$.

     Check for P3:   Suppose $\sigma \le \tau \le \rho$. If $\sigma = \tau$ or $\tau = \rho$, then $\sigma \le \rho$. Otherwise there are ν, η with $\widehat{\sigma \nu} = \tau$, $\widehat{\tau \eta} = \rho$. So $\rho = \widehat{\sigma \nu \eta}$, hence ρ extends σ, hence $\sigma \le \rho$.

Example 6. An extremely useful example of a partial order is the lexicographic (or dictionary) order on finite sequences. You have known this order since childhood: First there is an order on the letters of the alphabet, $a \le b \le c \ldots$. Given two words p and q, we say $p \le q$ if either $q = p$, q extends p (cat $\le$ cattle), or if, in the first place at which they differ, the letter occurring in q comes after the letter appearing in p (cacophony $\le$ cat).

We generalize this order as follows. For any set X, a useful set associated with X is the set of all nonempty finite sequences from X, which we call FIN(X). Now suppose X is a partially ordered set. Define the lexicographic order $\le_L$ on FIN(X) as follows: $\sigma \le \tau$ iff one of the following holds: $\sigma = \tau$, τ extends σ; or there are ρ, ν, η with $\sigma = \widehat{\rho \nu}$, $\tau = \widehat{\rho \eta}$, and the first element of ν is strictly less than the first element of η. We use the notation $L(X)$ to refer to the set FIN(X) under this lexicographic order.

For example, if $X = \mathbb{R}$, we have $(e, \pi) \le_L (\pi, e, \sqrt{2}) \le_L (\pi, \sqrt{11})$.

We show that if X is a partially ordered set, so also is $L(X)$. First some notation: $\le$ is the partial order on X and $\sigma(n)$ is the nth element of σ.

     Check for P1:   By definition each $\sigma \le_L \sigma$.

     Check for P2:   Suppose $\sigma \le_L \tau \le_L \sigma$ and $\sigma \ne \tau$. If τ extends σ, then inspecting the definition of $\le_L$ shows that $\tau \not\le_L \sigma$. So some n is the first place at which τ differs from σ. Then $\sigma(n) \leq \tau(n) \leq \sigma(n)$, so $\sigma(n) = \tau(n)$, which contradicts the definition of n. Hence $\sigma = \tau$.

Check for P3: Suppose $\sigma \leq_L \tau \leq_L \rho$. If either $\sigma = \tau$ or $\tau = \rho$, we are done. There are three remaining cases: (1) τ extends σ; (2) ρ extends τ; (3) neither of the previous cases holds.

We will complete case 3 and leave the other cases to the reader. Let k be the first place at which σ differs from τ; let n be the first place at which τ differs from ρ. If $n \geq k$, then $\rho(k) \geq \tau(k) \geq \sigma(k)$, and k is the first place at which $\rho(k) \neq \sigma(k)$, so $\rho \geq_L \sigma$. If $n < k$, then $\rho(n) \geq \tau(n) = \sigma(n)$, and n is the first place ρ differs from σ, so $\rho \geq_L \sigma$.

SECTION 1.2. SOME FACTS ABOUT PARTIALLY ORDERED SETS

Definition 7. Let $\leq$ be a partial order on a set X. Two elements x, y of X are comparable iff $x \leq y$ or $y \leq x$. They are compatible iff there is some z such that $z \leq x$ and $z \leq y$.

Note that if two elements are comparable then they are compatible, but not vice versa. In example 1, $3 \leq_D 6$ and $3 \leq_D 9$, but $6 \not\leq_D 9$ and $9 \not\leq_D 6$.

Definition 8. A subset B of a partially ordered set is a chain iff it is pairwise comparable; i.e., if $x, y \in B$, then x and y are comparable. B is an antichain iff no two elements of B are compatible. B is linked iff it is pairwise compatible.

Definition 9. A partial order $\leq$ on a set X is linear iff X is a chain, and we say X is a linearly ordered set, or linear.

For example, example 4 is a linear order, but examples 1, 2, 3, and 5 are not. Note that any subset of a linearly ordered set is linearly ordered. If X is not linear then $L(X)$ is not linear (you can find two one-element sequences that are not comparable), but on the other hand

Theorem 10. If X is linear, so is $L(X)$.

Proof. Suppose $\sigma, \tau \in L(X)$, where $\sigma \neq \tau$ and τ does not extend σ. Let k be the first place at which they differ. Then, since X is linear, either $\sigma(k) \leq \tau(k)$ or $\tau(k) \leq \sigma(k)$. So either $\sigma \leq_L \tau$ or $\tau \leq_L \sigma$.

Definition 11. Let X be a partial order, $x \in X$. We say

(a) x is minimal iff, for all $y \in X$, if $y \leq x$ then $y = x$.

(b) x is maximal iff, for all $y \in X$, if $y \geq x$ then $y = x$.

(c) x is a minimum iff, for all $y \in X$, $y \geq x$; i.e., x is the smallest element of X.

(d) x is a maximum iff, for all $y \in X$, $y \leq x$; i.e., x is the largest element of X.

Note that a minimum element is minimal and a maximum element is maximal.

Some examples: In example 5 the two one-element sequences 0 and 1 are minimal, and there are no maximal elements. In example 2 every element is both maximal and minimal. In example 3 if $X = \{y : y \subset \mathbb{N}\}$ then $\mathbb{N}$ is a maximum and $\emptyset$ is a minimum.

Proposition 12. A maximum is the unique maximal element. A minimum is the unique minimal element.

Proof. Suppose $y \leq x$ for all y. If y is a maximal element, then either x and y are not comparable, which is false, or $x \leq y$. So $x = y$. The proof for minimums is similar.

Note that a unique maximal element need not be a maximum and a unique minimal element need not be a minimum. For example, let $X = (0, 1] \cup [2, 3)$ under the ad hoc ordering $x \leq_A y$ iff $x, y \in (0, 1]$ and $x \leq y$ (where $\leq$ is the usual order on $\mathbb{R}$) or $x, y \in [2, 3)$ and $x \leq y$ (see Figure 1). Then 1 is the unique maximal element, but $1 \not\leq_A 2$; and 2 is the unique minimal element, but $2 \not\leq_A 1$.

Proposition 13. A maximal element in a linear order is a maximum. A minimal element in a linear order is a minimum.

Proof. Fix x. Since the order is linear, for every y in the order either $y \geq x$ or $y \leq x$. If x is minimal, then either $y \geq x$ or $y = x$ for all y, so x is a minimum. The proof for maximal elements is similar.

Figure 1

SECTION 1.3. EQUIVALENCE RELATIONS

Here are the axioms for an equivalence relation $\equiv$ on a set X:

For all $x, y, z \in X$,
E1 (Reflexive). $x \equiv x$.
E2 (Symmetric). $x \equiv y$ iff $y \equiv x$.
E3 (Transitive). if $x \equiv y$ and $y \equiv z$, then $x \equiv z$.

Example 14. Example 2 is an equivalence relation.

Example 15. Recall that for a set X we define $\text{FIN}(X)$ to be the collection of finite nonempty sequences of elements of X. We define two equivalence relations on $\text{FIN}(X)$: $\sigma \equiv_1 \tau$ iff σ and τ have the same length, and $\sigma \equiv_2 \tau$ iff the last element of σ equals the last element of τ. Thus, e.g., if $X = \mathbb{R}$, then $(\pi, e, \sqrt{2}) \equiv_1 (\sqrt{11}, 1, \pi/e)$ and $(\pi, e, \sqrt{2}) \equiv_2 (1, \sqrt{2})$. Both $\equiv_1$ and $\equiv_2$ are equivalent relations.

Example 16. Let F be the set of all functions whose domain is $\mathbb{N}$ and whose range is a subset of $\mathbb{N}$. For $f, g \in F$, define $f \equiv^* g$ iff $\{n: f(n) \neq g(n)\}$ is finite. Let us show that $\equiv^*$ is in fact an equivalence relation.

Check for E1: $\{n: f(n) \neq f(n)\} = \emptyset$ and hence is finite.
Check for E2: If $\{n: f(n) \neq g(n)\}$ is finite, so is $\{n: g(n) \neq f(n)\}$.
Check for E3: Suppose $\{n: f(n) \neq g(n)\}$ and $\{n: g(n) \neq h(n)\}$ are both finite. Suppose $f(n) \neq h(n)$ for some n. Then there are two possibilities: (Case 1) $f(n) = g(n)$ and $g(n) \neq h(n)$ and (Case 2) $f(n) \neq g(n)$. Note that there are only finitely many instances of each case, so $\{n: f(n) \neq h(n)\}$ is finite as we wished.

Definition 17. If x is an element of a set X with an equivalence relation $\equiv$, then the equivalence class of x is denoted $[x]$ and defined by $\{y \in X: y \equiv x\}$.

Notice that if $[y] = [x]$ and $y' \in [y]$, $x' \in [x]$ then $[y'] = [x']$.

Some notation: If $\equiv$ is an equivalence relation on X, we define $X/\equiv$ to be $\{[x]: x \in X\}$.

Example 18. We define a partial order on $F/\equiv^*$, as follows: $[f] \leq [g]$ iff $\{n: f(n) > g(n)\}$ is finite. This sort of definition should be familiar from analysis—we do not care what happens at the beginning of a sequence, just what happens from some point on.

We have to show that this is well-defined: If $f' \in [f]$ and $g' \in [g]$ and $[f] \leq [g]$, can we conclude that $[f'] \leq [g']$? If $[f] \leq [g]$, $f' \in [f]$, and $g' \in [g]$, then $\{n: f'(n) > g'(n)\}$ is a subset of the union of the following sets: $\{n: f'(n) \neq f(n)\}$, $\{n: f(n) > g(n)\}$, $\{n: g'(n) \neq g(n)\}$. Each of these sets is finite. So $[f'] \leq [g']$.

Proving that $\leq$ is a partial order on $F/\equiv^*$ is left as an exercise. This order is known as the Fréchet order, and its study has ramifications in many branches of mathematics.

There is a close relation between equivalence relations and partitions, where

Definition 19. A partition P of a set A is a collection of subsets of A so that every element of A belongs to exactly one element of P.

For example, if $a_p = \{n \in \mathbb{N}: p$ is the first prime number dividing $n\}$ for each prime p, then $P = \{\{0, 1\}\} \cup \{a_p: p$ a prime$\}$ is a partition of $\mathbb{N}$, since every natural number not equal to either 0 or 1 has a unique least prime dividing it.

Another example: If L_r is the horizontal line in the plane through the point $(0, r)$, then $P = \{L_r: r \in \mathbb{R}\}$ is a partition of the plane, since every point in the plane lies on a unique horizontal line and every horizontal line is some L_r.

Notice that the elements of a partition P of A are pairwise disjoint and that $A = \bigcup_{p \in P} p$, where $\bigcup_{p \in P} p = \{x:$ for some $p \in P$, $x \in p\}$.

We end this section by showing that partitions are essentially the same as equivalence relations.

If P is a collection of subsets of A, we write $x \, E_P \, y$ iff there is some $p \in P$ with $x, y \in p$.

Lemma 20. If P is a partition of A, then the relation E_P is an equivalence relation on A.

Proof. Notice that E_P is reflexive and symmetric for any P a collection of subsets of A. We show that E_P is transitive if P is a partition. Suppose $x, y \in p$ and $y, z \in q$, where $p, q \in P$. Since P is a partition, y is an element of exactly one element of P, so $p = q$ and $x \, E_P \, z$.

Definition 21. Given an equivalence relation E on a set A, we define $P_E = \{[x]: x \in A\}$, where $[x]$ is the equivalence class of x under E. P_E is usually written A/E, as in the remark after definition 17, or A mod E.

Lemma 22. If E is an equivalence relation on A, then P_E is a partition of A.

Proof. If $x \in [y]$ and $x \in [z]$, then $y E x E z$, so $y E z$, so $[y] = [z]$. Thus every element of A belongs to exactly one element of P_E, as required.

By working through the definitions, it is easy to show that $P_{(E_P)} = P$, for any partition P, and $E_{(P_E)} = E$ for any equivalence relation E.

SECTION 1.4. WELL-ORDERED SETS

In section 3.3 we will actually construct the mathematical universe (in a philosophically nonconstructive way). The spine of that universe will be the class of ordinals. Elsewhere we will do constructions and proofs using ordinals to mark the stages. When we talk about size we will usually insist that the infinite numbers marking size are ordinals. Although the universe will in some sense be built out of operations on the empty set, this approach is too nihilistic to be very useful. The fact that the universe is built out of operations on ordinals, however, is extremely useful. In short, to study set theory you must study ordinals. Whatever an ordinal is (we will not define it until Chapter 3), its purpose is to be a canonical well-ordered set—we will eventually prove that every well-ordered set is order-isomorphic to an ordinal. So to understand set theory we must understand ordinals, and to understand ordinals we must understand well-ordered sets.

Definition 23. A well-ordered set X is a linearly ordered set for which every nonempty subset has a minimal element.

Example 24. Since every nonempty set of natural numbers has a least element, $\mathbb{N}$ is well-ordered.

Example 25. Let $X = \{m/(m+1): m \in \mathbb{N}\} \cup \{1 + m/(m+1): m \in \mathbb{N}\}$ and order X by the usual order on the reals. Then X is well-ordered since it is the union of two increasing sequences and any nonempty subset must have either a least element in the first sequence or a least element in the second sequence.

Example 26. Let $X = \{m/(m+1): m \in \mathbb{N}\} \cup \{1 + m/(m+1): m \in \mathbb{N}\} \cup \{2 + m/(m+1): m \in \mathbb{N}\}$. Then X is well-ordered by the usual order on the reals. (The proof is left to the reader.)

Example 27. Let $X = \{n + m/(m+1): n, m \in \mathbb{N}\}$ ordered by the usual order on $\mathbb{R}$. Then X is well-ordered.

Proof. If Y is a nonempty subset of X then there is some least n such that $Y \cap \{n + m/(m+1): m \in \mathbb{N}\} \neq \emptyset$. Then there is some least m for which $n + m/(m+1) \in Y$. This is the minimal element of Y.

Being well-ordered is an example of a hereditary property, that is, a property inherited by every subset.

Proposition 28. Every subset of a well-ordered set is well-ordered.

Proof. If X is well-ordered, $Y \subset X$ and $Z \subset Y$, then $Z \subset X$, so Z has a minimal element or $Z = \emptyset$.

A basic criterion for well-ordered sets is the following theorem.

Theorem 29. A linear order X is well-ordered iff there is no infinite descending chain of elements of X.

Proof. An infinite descending chain has the form $x_0 > x_1 > x_2 > x_3 \ldots$. Suppose X has such a chain. Then $\{x_i: i \in \mathbb{N}\}$ has no least element, so X is not well-ordered. For the other direction suppose $Y \subset X$ and Y has no minimal element. Pick $x_0 \in Y$. Since x_0 is not minimal in Y, there is $x_1 \in Y$ with $x_1 < x_0$. Since x_1 is not minimal in Y, there is $x_2 \in Y$ with $x_2 < x_1$. And so on. Thus there is an infinite descending chain of elements of X.

Theorem 29 differs from the theorems we have seen so far in the sophistication of its proof. When we assumed that X was not well-ordered and constructed an infinite descending chain, no rule was given for choosing such x_i uniquely. Thus, given x_0 there may be many candidates for x_1, and given each of these there may be many candidates for x_2, and so on. That we can pick a path through this maze seems reasonable, but in fact we use a weak form of a somewhat controversial axiom, the axiom of choice. This axiom is defined and explored in chapter 3 and is critical to our discussion of size, or cardinality, in chapter 5. It is important to note later, when we work with well-orderings both with and without the axiom of choice, that only one direction of Theorem 29 holds without the axiom of choice.

An immediate corollary to Theorem 29 is that the closed unit interval $[0, 1]$ is not well-ordered. Neither is $[0, 1] \cap \mathbb{Q}$, where $\mathbb{Q}$ is the set of rational numbers.

Definition 30. Let X be a linear order and suppose $x \in X$. We say that y is the successor of x iff $y > x$ and, for all $z \in X$, if $z > x$ then $z \geq y$. We denote the successor of x by $S(x)$. We say that y is a limit if y is not the successor of any x.

Notice that x cannot have two distinct successors. Notice that not every element in a linear order need have a successor: For example, no element in $\mathbb{R}$ has a successor under the usual ordering. Notice that a minimal element is a limit. More examples: Every element in $\mathbb{R}$ is a limit under the usual ordering; and in Example 27 the limits are all n, where $n \in \mathbb{N}$, while the successors are all $n + m/(m+1)$, where $m \geq 1$.

Proposition 31. Let X be well-ordered, $x \in X$. Either x is maximal, or x has a successor.

Proof. If x is not maximal, $\{y : y > x\}$ is not empty. The minimal element of this set is $S(x)$.

All our examples of well-ordered sets looked like a copy of $\mathbb{N}$ followed by a copy of $\mathbb{N}$, followed by a copy of $\mathbb{N}$ .... In fact, Theorem 34 will show that this is essentially what all well-ordered sets look like (the "essentially" means that some well-ordered sets have a finite tail tacked on after all the copies of $\mathbb{N}$).

Definition 32. $S^0(x) = x$; $S^{n+1}(x) = S(S^n(x))$ for $n \geq 0$.

Proposition 33. If $n < m$, then $S^n(x) < S^m(x)$ for every x.

Proof. Suppose not. Let n_0 be some natural number so that there is some $m > n_0$ and some x for which $S^{n_0}(x) \geq S^m(x)$. For this n_0, let m_0 be the least natural number greater than n_0 so that $S^{n_0}(x) \geq S^{m_0}(x)$ for some x. Finally, let x_0 be an element so that $S^{n_0}(x_0) \geq S^{m_0}(x_0)$. Since $m_0 > n_0$. there is some k for which $m_0 = n_0 + k + 1$. By the definition of successor, k cannot equal 0. Hence, by hypothesis on m_0, $S^{n_0}(x_0) < S^{m_0-1}(x_0)$. By definition, $S^{m_0-1}(x_0) < S^{m_0}(x_0)$. So, by transitivity, $S^{n_0}(x_0) < S^{m_0}(x_0)$, which contradicts our definitions of m_0, x_0. Thus our hypothesis that the proposition fails is false.

Proposition 33 has a more straightforward proof using induction, as we will see in the next section.

Theorem 34. Let X be well-ordered, and let Λ be the set of limits in X.

Then

   (a) $X \subset \{S^n(x) : x \in \Lambda, n \in \mathbb{N}\}$.
   (b) If $x, y \in \Lambda$ and $x < y$, then $S^n(x) < y$.
   (c) If X has no maximum element, then neither does Λ.
   (d) If X has a maximum, then so does Λ. The maximum of X is some $S^n(x)$, where x is the maximum of Λ.

   Proof. (a) Suppose X is well-ordered and does not satisfy (a). Then $A = \{x \in X : x \neq S^n(z)$ for all $z \in \Lambda, n \in \mathbb{N}\}$ is nonempty. Let x be the minimum element of A. Then $x \notin \Lambda$, so $x = S^n(y)$ for some $y \in X$. Since $x \in A$, then $y \in A$. But then x is not minimal in A, which is a contradiction. So $X \subset \{S^n(x) : x \in \Lambda, n \in \mathbb{N}\}$.
   (b) Suppose not. Let n be least such that $S^n(x) \geq y$ or $S^n(x)$ is not defined. Then $n \neq 0$ since $y > x$. Since $n = m + 1$ for some m, either $y = S^n(x)$, or $S^n(x) \geq y > S^m(x)$; hence $y = S^n(x)$. But then $y \notin \Lambda$, a contradiction.
   (c) This follows immediately from (a).
   (d) Let y be the maximum of X. If $y \in \Lambda$, we are done. If not, then by (a), y is some $S^n(x)$, $x \in \Lambda$. By (b), x is the maximum of Λ.

   We close with a useful class of well-ordered sets.
   Let $L_n(X)$ be all sequences in $L(X)$ of length $\leq n$, under the order $\leq_L$.

Theorem 35. If X is well-ordered, so is each $L_n(X)$.

   Proof. Let Y be a nonempty subset of $L_n(X)$. Then there is some x_1 which is minimal for $\{\sigma(1) \in X : \sigma \in Y\}$. If $\sigma, \tau \in Y$ and $x_1 = \sigma(1)$ and $\tau(1) \neq x_1$, then $\sigma \leq_L \tau$. So we restrict our attention to $A_1 = \{\sigma : \sigma(1) = x_1\}$. If $(x_1) \in A_1$, we have found our minimal element and are done. If not, every sequence in A_1 has length at least 2. So let x_2 be minimal for $\{\sigma(2) : \sigma \in A_1\}$. If $\sigma, \tau \in A_1$ and $\sigma(2) = x_2$, then $\sigma \leq_L \tau$. So if $(x_1, x_2) \in Y$, we have found our minimal element in Y. If not, we restrict our attention to $A_2 = \{\sigma \in A_1 : \sigma(2) = x_2\}$ and note that every element of A_2 has length at least 3. And so on. Since every sequence in Y has length at most n, the process stops in k steps for some $k \leq n$. We then have $Y \supset A_1 \supset A_2 \supset \cdots \supset A_k$ and $x_1, x_2, \ldots, x_k$, where, if $\sigma, \tau \in Y$ and $\sigma(i) = x_i$ for $i = 1, \ldots, k$, then $\sigma \leq_L \tau$, and $(x_1, \ldots, x_k) \in Y$. But then $(x_1, \ldots, x_k)$ is the minimal element we want.

   Thus, for example, each $L_n(\mathbb{N})$ is well-ordered. The limits are the minimal element (0) and all words of length less than n not ending in 0.

Let us check this for $L_2(\mathbb{N})$. Each one-element sequence $(n+1)$ is a limit: Any sequence in $L_2(\mathbb{N})$ below $(n+1)$ is some two-element sequence (m, k), where $m \le n$; the successor of (m, k) is $(m, k+1)$; and $(m, k+1) < (n+1)$. On the other hand, no sequence of length 2 is a limit: Each $(m, k+1)$ is the successor of (m, k), and each $(m, 0)$ is the successor of (m).

Now for the general proof: $\widehat{\sigma(m+1)} > \widehat{\sigma(m)}\tau$ for all τ; the word $(x_1, \ldots, x_n)$ is the successor of $(x_1, \ldots, x_n - 1)$ if $x_n \ne 0$; and $(x_1, \ldots, x_k, 0)$ is the successor of $(x_1, \ldots, x_k)$.

It is a useful exercise to embed the $L_n(\mathbb{N})$'s into $\mathbb{R}$. For example, the reader can check that $L_2(\mathbb{N})$ is order-isomorphic to the set of Example 27. (We say that X is order-isomorphic to Y iff there is some 1–1 function f from X onto Y so that $x \le_X z$ iff $f(x) \le_Y f(z)$ for all $x, z \in X$. We say that X embeds in Y iff X is order-isomorphic to a subset of Y.)

It is not necessarily true that if X is well-ordered then so is $L(X)$. In exercise 19, the reader is invited to show that $L(\{0, 1\})$ is not well-ordered, where the order on $\{0, 1\}$ is $0 \le 1$.

SECTION 1.5. MATHEMATICAL INDUCTION

Recall the principle of mathematical induction:

Mathematical Induction, Version I. If 0 is in a set X and if "$n \in X$" implies "$n + 1 \in X$" for all natural numbers n, then every natural number is an element of X.

Two variations are

Mathematical Induction, Version II. If "$k \in X$ for all $k < n$" implies "$n \in X$," for every natural number n, then every natural number is an element of X.

Mathematical Induction, Version III. If $j \in X$ and "$n \in X$" implies "$n + 1 \in X$", for every $n \ge j$, then every natural number $n \ge j$ is an element of X.

Principles I and II are easily seen to be the same (just notice that 0 has no predecessors in $\mathbb{N}$; thus the hypothesis of II vacuously ensures that $0 \in X$). Principle III is just principle I moved up a bit.

Notice that the principle of mathematical induction follows from the fact that $\mathbb{N}$ is well-ordered: If "$k \in X$ for all $k < n$" implies "$n \in X$", for every $n \in \mathbb{N}$, and if $\mathbb{N} - X \ne \emptyset$, then let n be the least element of $\mathbb{N} - X$. But $k \in X$ for all $k < n$ by hypothesis on n. So, by the induction hypothesis, $n \in X$, which contradicts the definition of n.

You have probably seen induction used to prove statements about the natural numbers, for example that $\sum_{i=1}^{n} i = n(n+1)/2$. We give three examples of how to use induction. The first example assumes naive set theory.

Let $\mathscr{P}(X)$, called the power set of X, be the set of subsets of a set X.

Theorem 36. If X is finite, so is $\mathscr{P}(X)$.

Proof. Suppose we know that for every set x of size less than n, $\mathscr{P}(x)$ is finite. Let z be a set of size n. If z has no element, we have nothing to prove since it has only one subset, namely itself. So we may suppose there is some element $a \in z$. Then every subset y of z falls into two classes: $y \subset z - \{a\}$ or $y = x \cup \{a\}$, where $x \subset z - \{a\}$. By the induction hypothesis there are only finitely many sets in the first class and hence only finitely many in the second class. The sum of two finite numbers is finite, so we are done.

The second example gives an alternative proof for the last theorem in section 1.4.

Theorem 35. If X is well-ordered, so is each $L_n(X)$.

Proof. If $n = 0$, there is nothing to prove. So suppose $n > 0$ and suppose $L_{n-1}(X)$ is well-ordered. For each σ of length n, let σ^* be the first $n-1$ elements of σ, and let $\sigma(n)$ be the nth element of σ. If $\{\sigma_i : i \in \mathbb{N}\}$ is a descending chain in $L_n(X)$, then infinitely many elements of the chain have the same length k. By hypothesis k cannot be smaller than n, so without loss of generality each σ_i has length n. Since each $\sigma_{i+1}* \leq \sigma_{i}*$, by the induction hypothesis there is some j so that if i, $i' \geq j$ then $\sigma_{i}* = \sigma_{i'}*$. Hence $\sigma_{i+1}(n) < \sigma_i(n)$ for all $i \geq j$, which contradicts X being well-ordered.

We have just shown that $L_0(X)$ is well-ordered and, for all $n > 0$, "$L_{n-1}(X)$ is well-ordered" implies "$L_n(X)$ is well-ordered." Hence, by the principle of induction I, each $L_n(X)$ is well-ordered, and we are done.

A third example is an inductive proof of proposition 33. Recall

Proposition 33. If $n < m$, then $S^n(x) < S^m(x)$ for every x.

Proof. Fix n. We will show by induction on m that the statement holds for all $m > n$. By the definition of successor, it holds if $m = n + 1$.

Now suppose it holds for m. We must show it holds for $m + 1$. For each x, $S^n(x) < S^m(x)$ by the induction hypothesis, and $S^m(x) < S^{m+1}(x)$ by the definition of a successor. So by transitivity, $S^n(x) < S^{m+1}(x)$, and we are done.

A concept related to proof by induction is that of a recursive construction in which we build something up in stages: What we have at stage n determines what we do at stage $n + 1$. Here is a simple example.

Theorem 37. Let $\{a_n: n \in \mathbb{N}\}$ be a sequence of infinite subsets of $\mathbb{N}$ so that each $a_{n+1} \subset a_n$. Then there is an infinite set a so that $a - a_n$ is finite for each n; i.e., for each n, all but finitely many elements of a are elements of a_n.

Proof. We construct $a = \{k_1, k_2, \ldots\}$ recursively. The requirement at stage n is that $k_m \in a_m$ for all $m \leq n$. (Note: This sort of requirement serves the same function as the inductive hypothesis in a proof by induction.) Suppose we have constructed $k_1, \ldots, k_n$. Since a_{n+1} is infinite, it has some element k in it which equals no k_i, for $i \leq n$. Let k_{n+1} be such a k, and keeping going.

Let us prove that this construction works. Since $n < m$ implies $a_n \supset a_m$, if $n < m$ then $k_m \in a_n$. Thus each a_n contains all but finitely many elements of a.

Let us analyze the notion of a recursive construction on the natural numbers, using theorem 37 as a guide. A recursive construction on the natural numbers is a sequence of sets $\{A_n: n \in \mathbb{N}\}$ (in our example, $A_n = \{k_i: i \leq n\}$), where $A_n \subset A_{n+1}$ for each n, and the final set constructed is $A = \bigcup_{n \in \mathbb{N}} A_n$.

There is at least one requirement which holds for each A_n (in theorem 37 the requirement is that $k_m \in a_m$ for all $k_m \in A_n$), and this requirement is used to prove that A has the desired property. Notice that the phrase "recursive construction on the natural numbers" does not mean that A is a subset of $\mathbb{N}$ but rather that the elements of $\mathbb{N}$ are used to index the approximations of A.

We will not give the proof that recursive constructions work, since the general theorem is abstract enough to lose coherence for many students (the interested student can find a proof in, say, Jech's *Set Theory*). There are two points to prove: The set A exists, and A does what it is supposed to do. Existence comes about via the axiom of replacement (see section

2.9); that A works usually involves induction. The interested student can check our particular recursive constructions against these principles.

Exercises using induction and recursion can be found from chapter 2 on.

SECTION 1.6. MODELS

In this section we look at earlier material from the point of view of mathematical logic. There are three types of objects to be considered: languages, theories, and models.

First-Order Languages

In sections 1.1 and 1.3, we introduced sets of axioms for partially ordered sets and for equivalence relations. These axioms were expressed in extremely restricted languages, languages with the following properties:

(a) There was a fixed set of symbols for relations—"$\leq$" in section 1.1 and "$\equiv$" in section 1.2.

(b) All sentences of the language were made up of logical words or symbols ("and," "if ... then," "for all," "=," and so on), the symbols of (a), and variables.

(c) There was only one kind of object referred to (in our examples the only objects considered were the points of the underlying set X).

With some minor augmentation to allow for functions and constants, and more precision to rule out ambiguities, properties (a), (b), and (c) define what are known as first-order languages.

Languages with properties (a) and (b) belong to the class of languages known as formal languages. Property (c) is the special property of first-order languages. For example, the statement "every subset of a partially ordered set is partially ordered in the restricted ordering" cannot be put in the first-order language of partially ordered sets—we cannot say "subset." Another example is the statement "X is well-ordered by $\leq$." The definition of well-ordered is that every subset has a certain property. But again, "subset" is not in our vocabulary.

An example of a statement that can be translated into a first-order language is

$$\text{"}X \text{ is linearly ordered by } \leq \text{"}$$

whose translation is

$$"\forall x \, \forall y \, \forall z[x \leq x \text{ and (if } x \leq y \text{ and } y \leq x \text{ then } y = x) \text{ and}$$
$$(\text{if } x \leq y \text{ and } y \leq z \text{ then } x \leq z) \text{ and } (x \leq y \text{ or } y \leq x)]"$$

Here the first three clauses tell us that X is partially ordered; the last tells us that X is a chain. Notice that we do not actually refer to X in the statement.

In Section 2.1 we will define the first-order language of set theory.

Theories

A first-order language does not, in itself, have any meaning. Just because we use the symbol "$\leq$" does not mean that we have a partial order; just because we use the symbol "$\equiv$" does not mean that we have an equivalence relation. We give meaning to symbols by adopting a set of axioms: P1 through P3 for $\leq$ and E1 through E3 for $\equiv$. Two types of theorems can be proved from a set of axioms: theorems in the language (e.g., propositions 12 and 13) and theorems that reach outside the language (e.g., theorem 10). It is worthwhile to restate proposition 12, first in English, and then in the restricted language.

"A maximum is the unique maximal element. A minimum
is the unique minimal element."

becomes

$$"\forall x[(\forall y \, y \leq x) \to \forall y((\forall z \, z \not\geq y) \to y = x)] \text{ and}$$
$$\forall x[(\forall y \, y \geq x) \to \forall y((\forall z \, z \not\leq y) \to y = x)]."$$

The reader is invited to translate proposition 13 into the first-order language of partial orders.

As for theorem 10—if X is linear, so is $L(X)$—it starts from X and proves something about $L(X)$; but there is no way to refer to $L(X)$ or even its elements in the first-order language of partial orders.

We must be careful; things are not always as they seem. Thus lemma 22 ("if E is an equivalence relation on A, then P_E is a partition of A"), even though it is not first-order, translates into a first-order theorem of equivalence relations; to wit: "if $\forall x \forall y \forall z[x \equiv x \text{ and } (x \equiv y \text{ iff } y \equiv x) \text{ and}$ (if $x \equiv y$ and $y \equiv z$ then $x \equiv z)]$ then $[\forall x \exists y(y \equiv x) \text{ and } \forall x \forall y \forall z$ (if $z \equiv x$ and $z \equiv y$ then $y \equiv x)]$."

A first-order theory is the collection of first-order formulas which can be proved from a set of axioms. Thus the collection of sentences in the language of partial orders provable from P1 through P3 is called the first-order theory of partial orders. Proposition 12 is in this theory, but theorem 10 is not.

Not every important theory is equivalent to a first-order theory. For example, it can be shown that the concept of well-ordering cannot be captured by a first-order theory.

Models

Roughly speaking, a model of a theory is a structure for which the theory is true. Thus a model of the theory of partial orders is just a partially ordered set; a model of the theory of equivalence classes is just a set with an equivalence relation. In algebra, a model of group theory is a group and a model of ring theory is a ring. And so on. The study of models of first-order theories is called model theory. Using the language of model theory, in the examples of sections 1.1 and 1.3 we were verifying whether certain structures were models of certain theories.

There are two salient facts about checking whether a structure is a model of a first-order theory.

(A) You have to check only whether the axioms hold.

Thus once a structure satisfies P1, P2, and P3, it automatically satisfies propositions 12 and 13.

(B) You never look outside the structure.

Thus in checking example 1, recall that $2 = (2/3)(3)$ did not imply $2 \leq_D 3$ since 2/3 was not an element of the structure.

A theory is said to consistent iff it contains no contradictions. Consistent theories are, of course, the only ones worth studying.

Here are four important theorems about consistent first-order theories.

1. *The Completeness Theorem, Version I.* A first-order statement ϕ can be proved from a set of first-order axioms $\mathcal{A}$ iff ϕ holds in all models of $\mathcal{A}$. (Note that this is essentially principle (A) above.)
2. *The Completeness Theorem, Version II.* A first-order theory is consistent iff it has a model. (The interested reader can prove this equivalent to version I.)

3. *The First Incompleteness Theorem.* If a consistent, axiomatizable[†] theory is sufficiently complicated (in particular, if it encodes simple arithmetic on the natural numbers), then there is a statement in the language of the theory which the theory can neither prove nor disprove.

4. *The Second Incompleteness Theorem.* If a consistent, axiomatizable theory is sufficiently complicated (in particular, if it encodes simple arithmetic on the natural numbers), then it cannot prove its own consistency.

All four of these theorems are due to Gödel.

Consequences for Set Theory

These concepts—of first-order theory and of model—have had profound effects on set theory. Set theory—in which we claim to be able to embed all of mathematics—is a first-order theory. Thus the completeness theorems apply. Furthermore, if it embeds all of mathematics, it certainly embeds simple arithmetic (we will actually do some of this in chapter 4), so the incompleteness theorems apply. It is the applications of the completeness and incompleteness theorems that give modern set theory much of its power and beauty.

One can made a case for the statement that modern set theory is largely the study of models of set theory. Certainly it is the incompleteness theorems that have given set theory such essential application in other branches of mathematics. These themes—models and incompleteness—will reappear throughout the book.

A word on what we will mean by models. The precise, formal definition of "model of a first-order theory" is quite abstract, and far too general to be stated in a set-theory text. Instead we will take the ad hoc approach of defining what it is for a set to be a model of each axiom as the axiom is introduced, always appealing to principle (B) above. The models we define are known as standard models. As the axioms get more complicated, the definitions of their models become more complicated. When this happens, we will simplify life by only defining what it is for a certain kind of set (a transitive set) to be a model.

[†]An axiomatizable theory is, roughly speaking, one with a recognizable set of axioms, e.g., the theory of partial orders, or set theory. The definition of "recognizable" is that there is an effective algorithm which can decide whether an arbitrary formula in the language is an axiom or not. An effective algorithm is one which can be run on a computer. (More formal definitions exist, but this captures our intuiton well.)

The astute reader may have noted a seeming conflict between Gödel's second incompleteness theorem, which implies that set theory cannot prove that it has models, and the statement that modern set theory is largely the study of models of set theory. Both are true. Doing set theory is an act of faith, the same act of faith performed when we do any moderately complex mathematics, and the faith is that what we are doing is internally consistent. Formally, theorems involving consistency and models of set theory usually have one of two forms: "if set theory is consistent, then...," and "a model built in this way has those properties," thus escaping conflict with the second incompleteness theorem.

EXERCISES FOR CHAPTER 1

1. Show that the relation $x \leq_S y$ in example 3 is a partial order.

2. Define $x \leq_m y$, for $x, y \in \mathbb{N}$, iff $y - z = x$ for some z in $\mathbb{N}$. Is this a partial order?

3. Define the dual R^{-1} of a relation R by $x R^{-1} y$ iff $y R x$. Show that R is a partial order iff R^{-1} is.

4. Show that the relation $x E y$ defined by "$x E y$ iff $x = y$ or x is an element of y" is not a partial order.

5. Let X consist of all points (a, b) in the plane where $a, b \in \mathbb{N}$. Show that under the order "$x \leq y$ iff x sits directly below y" X has infinitely many chains, infinitely many minimal elements, and no maximal elements. What about $\leq^{-1}$? (See exercise 3).

6. Let Y consist of all points (a, b) in the plane where a, b are in $\mathbb{Z}$ and define "$x \leq y$ iff $x = y$ or x sits directly below y." How many chains does Y have? How many minimal elements? How many maximal elements?

7. Prove that a maximal element in a linear order is a maximum.

8. Find a partial order with seven maximal elements and three minimal elements.

9. Let X be as in exercise 6. Show that for fixed b, $\{(a, b): a \in \mathbb{N}\}$ is an antichain. More generally, show that A is an antichain in X iff no two elements of A have the same first coordinate.

10. Show that example 2 (see example 14) is an equivalence relation.

11. Consider the unit square $I^2 = [0, 1] \times [0, 1]$ in $\mathbb{R}^2$. For $(a, b), (c, d) \in I^2$, write $(a, b) E (c, d)$ iff $a = c$ and ($b = 1 - d$ or $b = d$). Show that this is an equivalence relation. If we glue y to x iff $x E y$ what geometrical shape results?

12. Again on the unit square, define $(a, b) F (c, d)$ iff $a = c$ and $b = d$; or $a = c$, $bd = 0$, and $b = 1 - d$. Show this is an equivalence relation. If we glue x to y iff $x F y$, what geometrical shape results?

13. Show that the Fréchet order (see example 18) is a partial order.

14. Consider the point $(1/3, 2/3)$ in the unit square. Referring to exercise 11, what is the equivalence class of this point? Referring to exercise 12, what is its equivalence class?

15. Draw a picture of $L_3(\mathbb{N})$; of $L_4(\mathbb{N})$. How many limits does each have?

16. Find a well-ordered subset of $\mathbb{R}$ isomorphic as an ordering to $L_3(\mathbb{N})$.

17. Find a well-ordered subset of $\mathbb{R}$ with exactly seven limit elements and one maximal element, which is also a limit.

18. Find a well-ordered subset of $\mathbb{R}$ with exactly seven limit elements and no maximal elements.

19. (a) Let $\{0, 1\}$ have the order $0 \leq 1$. Show that this is well-ordered but $L(\{0, 1\})$ is not well-ordered.
(b) Show that $L(X)$ is not well-ordered if the well-ordered set X has at least two elements.

20. Suppose X is partially ordered and every nonempty subset of X has a least element. Show that X is well-ordered.

21. Let X be well-ordered by $\leq_X$, and let Y be well-ordered by $\leq_Y$. Let $F = \{f : f$ is a function from X to $Y\}$. Let $\leq_L$ be defined on F by $f <_L g$ iff, where $x = \min\{y \in X : f(y) \neq g(y)\}$, $f(x) \leq_Y g(x)$. Show that F is linearly ordered by $\leq_L$.

2

THE AXIOMS, PART I

INTRODUCTION

In the nineteenth and early twentieth centuries mathematicians and philosophers were concerned with the foundations of mathematics far more urgently than they are today. Beginning with the effort to free calculus from its reliance on the then half-mystical concept of infinitesimals (there is a way of making infinitesimals precise, but it is modern, an unexpected fallout from formal mathematical logic), mathematicians interested in foundations were largely concerned with two matters—understanding infinity and exposing the basic principles of mathematical reasoning. The former will be dealt with later; it is with the latter that we concern ourselves now.

The basic principles of mathematical reasoning with which we will be concerned are the axioms of set theory. (There are other basic matters to deal with—for example, the laws of logic that enable us to discriminate a proof from a nonproof—which are the subject of mathematical logic.) Even restricting ourselves to deciding which statements are obviously true of the universe of sets, we find ourselves with several axiom systems to choose from. Luckily, the ones in common mathematical use are all equivalent (that is, they all prove exactly the same first-order theorems), as they should be if, indeed, they codify our common intuition. The system we shall use is called ZF, for the mathematicians Zermelo and Fraenkel, and is the one most widely used.

Why should we bother with axioms at all? Isn't our naive notion of a set enough? The answer is no. Promiscuous use of the word "set" can get us into deep trouble. Consider Russell's paradox: Is the set of all sets which are not members of themselves a set? If X is defined by $x \in X$ iff $x \notin x$, then $X \in X$ iff $X \notin X$, which is a contradiction. No mathematical theory which leads to a contradiction is worth studying. And if we claim that we can embed all of mathematics within set theory, as we shall, the presence of contradictions would call into question the rest of mathematics.

Another reason for elucidating axioms is to be precise about what is allowed in mathematical arguments. For example, in theorem 29 of chapter 1, we constructed a set by finding a first element, then a second, then a third, and so on . . . , and then gathering everything together "at the end." Is this a reasonable sort of procedure? We probably do not want to appeal to logic alone to justify it, since most people prefer proofs to be finite objects, and we are talking about an infinite process. The axioms of set theory will help us here. They will show that some infinite arguments are clearly sound (e.g., induction) and will also set limits on what we can do—the axiom of choice is controversial precisely because it allows a form of argument which some mathematicians find too lenient.

By the completeness theorem, the existence of a set of first-order axioms which defines set theory would carry with it the benefit of the existence of models of set theory, if we could show the consistency of set theory. But, by the second incompleteness theorem, we cannot show this from within ZF. The consistency of ZF and the existence of models of ZF are articles of faith.

If ZF is consistent, then, by the first incompleteness theorem, it has statements which hold in some models but not in others, just as some partial orders are linear and others are not. Such statements are called *independent*. Their associated questions (e.g., "is ϕ true?") are called *undecidable*. An easily stated undecidable question in ZF is: How many real numbers are there? Since all of mathematics can be done in the context of set theory, it is not unreasonable to expect that there are independent statements of general mathematical interest, and in fact there are. The most fertile fields for them so far are topology, algebra, and analysis.

A note on how this chapter is written: By the nature of the subject, there are times when we may seem unduly picky. Thus we must justify the use of ordered pairs, show that Cartesian products exist, and so on. Since we are trying to develop set theory axiomatically, ideally at each point we would use only the little bit of mathematics that has been deduced so far. But many of the concepts are best illustrated with reference to examples which can only be justified later, when we know more, but which are intuitively clear, the sort of thing the reader is mathematically used to. To avoid confusion, in this chapter and chapter 3 when we refer to concepts that cannot yet be justified by what we have done so far, the text will be marked off by indentation.

SECTION 2.1. THE LANGUAGE, SOME FINITE OPERATIONS, AND EXTENSIONALITY

Any mathematical theory must begin with undefined concepts or it will end in an infinite regress of self-justification. For us the concepts are the noun "set" and verb form "is an element of." We have an intuition of their meanings. The axioms are meant to be a precise manifestation of as much intuition as possible.

We write "$x \in y$" for "x is an element of y"; "$x \notin y$" for "x is not an element of y." Note that in the formal mathematical language the word "set" is superfluous—all the objects we talk about are sets. So we really have only the undefined symbol "$\in$" and the logical shorthand "$\notin$."

Using just "$\in$" and "$\notin$" we define some basic concepts.

Definition 1

(a) $x \subset y$ iff $\forall z$(if $z \in x$ then $z \in y$).
(b) $x \cup y = z$ iff $\forall w(w \in z$ iff $(w \in x$ or $w \in y))$.
(c) $x \cap y = z$ iff $\forall w(w \in z$ iff $(w \in x$ and $w \in y))$.
(d) $x - y = z$ iff $\forall w(w \in z$ iff $(w \in x$ and $w \notin y))$.
(e) $z = \emptyset$ iff $\forall w(w \notin z)$.

These new symbols are, strictly speaking, unnecessary. For example, we will shortly prove

$$\forall x \forall y(\text{if } x \subset y \text{ and } y \subset x \text{ then } x = y)$$

But this is just shorthand for

$$\forall x \forall y[\text{if}(\forall z(\text{if } z \in x \text{ then } z \in y) \text{ and}$$
$$\forall z(\text{if } z \in y \text{ then } z \in x)) \text{ then } x = y].$$

Definition 1 is not quite formal enough—it uses English words such as "or," "and," "if," "then." This can be avoided by restricting the language of set theory as follows:

We allow variable symbols: $x_0, x_1, x_2, \ldots$.

We allow constant symbols: $c_0, c_1, c_2, \ldots$[†]

We allow other logical symbols:

$\forall$ (for all)

$\exists$ (there exists)

$\wedge$ (and)

$\vee$ (or)

$\rightarrow$ (if . . . then . . .)

$\leftrightarrow$ (iff)

$=$ (equals)

$\neg$ (not)

((left parenthesis)

)   (right parenthesis)

We allow a single nonlogical symbol: $\in$. That is all.

Thus, for example, we rewrite

$$\forall w \in z (w \in x \text{ and } w \notin y)$$

as

$$\forall x_1 ((x_1 \in x_2) \rightarrow (x_1 \in x_3 \wedge \neg(x_1 \in x_4))).$$

Nearly everything we do in set theory can be done in this formal language, but our human brains find complicated formulas difficult to process.

Later we will define, in the language of set theory, an important class of objects known as ordinals. Let us see how an intuitively clear statement about ordinals translates into the formal language. Compare the statement:

if x, y are ordinals, then either $x \in y$ or $y \in x$ or $x = y$

with its formal equivalent:

[†]Constant symbols are used much as proper names are used, to refer to specific objects in a specific context (much as there are many Judys in the world, yet the name "Judy" causes no ambiguity when it is understood to refer to me). Thus the sentence "$\forall x_1(x_1 \notin c_3)$" changes its truth value according to which set is currently being named by c_3, just as the question "Is Judy a mathematician?" has a different answer according to which Judy is being referred to.

$$\forall x \forall y ((\forall z \forall w ((z \in w \wedge w \in x) \rightarrow z \in x)$$
$$\wedge \forall z \forall w ((z \in w \wedge w \in y) \rightarrow z \in y)$$
$$\wedge \forall z \forall w ((z \in x \wedge w \in x) \rightarrow (z \in w \vee z = w \vee w \in z))$$
$$\wedge \forall z \forall w ((z \in y \wedge w \in y) \rightarrow (z \in w \vee z = w \vee w \in z)))$$
$$\rightarrow (x \in y \vee x = y \vee y \in x)).$$

Even this is not formal enough—we should change x to x_1, y to x_2, z to x_3, w to x_4. But this is unreadable. Being only human, we feel free to use our abbreviations, that is, our defined symbols and concepts, and to use the English language.

Having accepted the symbols of definition 1, we would like to prove theorems about the operations they represent. To do this we need our first axiom, meant to capture our intuition that a set is defined by its elements, not by its description. For example, $\{x \in \mathbb{N}: x$ is divisible by $2\} = \{x \in \mathbb{N}: 3x$ is divisible by $6\}$.

Axiom of Extensionality. Two sets are equal iff they have the same elements: $\forall x \forall y [(x = y) \leftrightarrow \forall z (z \in x \leftrightarrow z \in y)]$.

Theorem 2. For all x and y, $x = y$ iff ($x \subset y$ and $y \subset x$).

Proof. Given x, y,

$$x = y \text{ iff } \forall z (z \in x \leftrightarrow z \in y)$$
$$\text{iff } \forall z ([z \in x \rightarrow z \in y] \text{ and } [z \in y \rightarrow z \in x])$$
$$\text{iff } [\forall z (z \in x \rightarrow z \in y) \text{ and } \forall z (z \in y \rightarrow z \in x)]$$
$$\text{iff } (x \subset y \text{ and } y \subset x).$$

Theorem 2 provides us with a method of proof. To show that two sets are equal, it suffices to show that each is a subset of the other.

Theorem 3. Let x, y, z be sets. Then

(a) $x \cup x = x = x \cap x$ (idempotence of $\cup$ and $\cap$).
(b) $x \cup y = y \cup x$; $x \cap y = y \cap x$ (commutativity of $\cup$ and $\cap$).
(c) $x \cup (y \cup z) = (x \cup y) \cup z$; $x \cap (y \cap z) = (x \cap y) \cap z$
(associativity of $\cup$, $\cap$).

(d) $x \cup (y \cap z) = (x \cup y) \cap (x \cup z)$; $x \cap (y \cup z) = (x \cap y) \cup (x \cap z)$
$$\text{(distributive laws)}.$$

(e) $x \subset y$ iff $x \cap y = x$ iff $x \cup y = y$.

(f) $x - (y \cup z) = (x - y) \cap (x - z)$; $x - (y \cap z) = (x - y) \cup (x - z)$
$$\text{(De Morgan's laws)}.$$

(g) $\emptyset \subset x$.

Proof. We content ourselves with proving (e) and (g), leaving the rest as an exercise. For (e): Suppose $x \subset y$. If $z \in x$, then $z \in x \cap y$. Hence $x \subset x \cap y \subset x$, so $x \cap y = x$. Suppose $x \cap y = x$. If $z \in x$, then $z \in y$ by hypothesis. Hence $y \subset x \cup y \subset y$, so $x \cup y = y$. Finally, if $x \cup y = y$ and $z \in x$, then $z \in x \cup y$, so $z \in y$. Hence $x \subset y$. Having come full circle, (e) is proven.

For (g): Let x be a set. We must show that $\forall z (z \in \emptyset \rightarrow z \in x)$. But "$z \in \emptyset$" is false for all z; hence the implication we need is vacuously true.

Let us define some more notation. If $x_1 \ldots x_n$ are sets, then $\{x_1, \ldots, x_n\} = z$ iff $(x_1 \in z \wedge \cdots \wedge x_n \in z) \wedge \forall y \in z(x_1 = y \vee \cdots \vee x_n = y)$. If ϕ is a formula in the language of set theory, then $\{x: \phi(x)\} = z$ iff $\forall y (y \in z \leftrightarrow \phi(y))$. For example, $\{x: x \in \mathbb{N} \text{ and } x > 10\} = \{11, 12, 13 \ldots\}$. Curly brackets will also be used less formally, as in $\{2, 4, 6, 8, 10 \ldots\}$ or $\{x_1, x_2, x_3, x_4 \ldots\}$, or more generally $\{x_i : i \in I\}$. These last uses will be formally defined in section 2.5, but we need them earlier for examples.

It is time now to define the sets we are willing to call models of the axiom of extensionality.

Suppose X is a set and x, y are distinct elements of X. Descending for a moment into anthropomorphic language, when can X recognize that $x \neq y$? Not only must there be some $z \in (x - y) \cup (y - x)$, but such a z must be an element of X—otherwise X is blind to its existence.

(An analogy: $\mathbb{N}$ does not know of the existence of any multiplicative inverses other than 1; but $\mathbb{Q}$ does.)

Definition 4. A set X satisfies (is a model of) extensionality iff, for all distinct $x, y \in X$, $X \cap [(x - y) \cup (y - x)] \neq \emptyset$.

Let us look at an example. $X = \{\emptyset, \{\emptyset\}, \{\{\emptyset\}\}, \ldots\}$. That is, if $x_0 = \emptyset$ and, for each $n \in \mathbb{N}$, $x_{n+1} = \{x_n\}$, then $X = \{x_n : n \in \mathbb{N}\}$. By extensionality and a little induction, $x_n \neq x_m$ iff $n \neq m$. Suppose $n < m$. Then $x_{m-1} \in x_m - x_n$. So X satisfies extensionality.

Another example: $X = \{x, y\}$ where $X \cap [(x - y) \cup (y - x)] = \emptyset$ and $x \neq y$. Then X does not satisfy extensionality, since any set which would enable it to distinguish x from y is excluded.

SECTION 2.2. PAIRS

We already defined $\{x, y\}$ in section 2.1. Our second axiom justifies talking about it.

Pairing Axiom. For all sets $x, y, \{x, y\}$ is a set: $\forall x \, \forall y \, \exists z \, (z = \{x, y\})$.

Note that $\{x, x\} = \{x\}$ by the axiom of extensionality. We call $\{x\}$ a *singleton* and $\{x, y\}$, where $x \neq y$, an *unordered pair* or, when the context is clear, just a pair. As we are about to see, the pairing axiom is actually quite strong.

Often in mathematics we need to distinguish between two objects in a pair. For example, the point $(2, 1)$ in $\mathbb{R}^2$ does not equal the point $(1, 2)$ in $\mathbb{R}^2$. We need to define a concept of ordered pair (x, y) that will have the property: $(x, y) = (z, w)$ iff $x = z$ and $y = w$. This definition must be reducible to a formula using only $\in$. There are many ways of doing this. The one used here is standard.

Definition 5. For all sets $x, y, (x, y) = \{\{x\}, \{x, y\}\}$.

When the context is unambiguous, we will take the liberty of calling an ordered pair just a pair.

Theorem 6. For all sets $x, y, z, w, (x, y) = (z, w)$ iff $x = z$ and $y = w$.

Proof. By extensionality and definition 5, if $x = z$ and $y = w$, then $(x, y) = (z, w)$.

For the other direction, suppose $(x, y) = (z, w)$. If $\{x\} = \{z, w\}$, then $x = w = z$ and $\{x, y\} = \{z\}$, so $x = y = z$ and $y = w$. If, on the other hand, $\{x\} = \{z\} \neq \{z, w\}$, then $x = z \neq w$. Since $\{x, y\} = \{z, w\}$, we again have $y = w$.

Ordered pairs enable us to define ordered n-tuples and finite dimensional Cartesian products, e.g., $\mathbb{R}^n$. There are two ways of doing this: an inductive method using minimal machinery, which works only for finite dimension, and a method using the definition of function, which generalizes to infinite dimension. We will eventually settle on the second method, but it is worthwhile to describe the first.

A Method for Cartesian Product

Define $(a) = \{a\}$, and, if $n > 2$, define $(a_0, \ldots, a_n) = ((a_0, \ldots, a_{n-1}), a_n)$. Define $X_0 \times \cdots \times X_n = \{(a_0, \ldots, a_n): a_i \in X_i \text{ for all } 0 \le i \le n\}$.

Thus, as an immediate corollary of the pairing axiom, if $x_0, \ldots, x_n$ are sets, so is $(x_0, \ldots, x_n)$. That $X_0 \times \cdots \times X_n$ is a set if each X_i is a set will be shown later.

The clumsiness of this definition is offset by the fact that it needs no further set-theoretic machinery. In fact it essentially does not need set theory at all—any theory with a definition of ordered pair satisfying theorem 6 will do. The classical axioms of number theory (Peano's axioms) give another example of such a theory.

We want to define models of the pairing axiom. If $x, y \in X$, $a \notin X$, and $\{x, y, a\} \in X$, then X thinks that $\{x, y, a\}$ satisfies the definition of $\{x, y\}$, since a is invisible to X. With this motivation we have

Definition 7. A set X satisfies (is a model of) pairing iff, for all $x, y \in X$, there is some $z \in X$ with $z \cap X = \{x, y\}$.

In particular, if, for all $x, y \in X$, $\{x, y\} \in X$—we say such an X is closed under pairing—then X satisfies pairing. However, this is not a necessary criterion.

Note that you can satisfy pairing and not satisfy extensionality, and vice versa. In general, a model for one axiom of ZF need not satisfy any of the others.

This simple process—unordered pair to ordered pair to arbitrary n-tuple—coupled with a little arithmetic and a lot of hard work gives us, in fact, Gödel's incompleteness theorems.

The first step is to code set theory by arithmetic. Thus, recalling the formal symbols of our language from Section 2.1, we code them as follows:

$\forall$	is assigned	0
$\exists$		1
$\neg$		2
$\wedge$		3
$\vee$		4
$\rightarrow$		5
$\leftrightarrow$		6
(		7
)		8

$=$	is assigned	9
$\in$		10
x_i		$11 + 2i$
c_i		$12 + 2i$.

Thus the formula $\forall x_1 \exists x_{17} \neg (x_1 = x_{17} \vee x_7 \in x_{17})$ is coded by the 14-tuple $(0, 13, 1, 45, 2, 7, 13, 9, 45, 4, 25, 10, 45, 8)$.

The second step, which will be done in the second and fourth chapters of this book, is to embed arithmetic inside set theory.

The third step (the interested reader can see proofs in any good logic text) is to show that, using arithmetic, there are formulas defining which sequences code formulas, proofs, theorems, and so on. For example, there is a set-theoretic formula ψ so that $\psi(x)$ holds iff x codes a theorem of set theory. The statement "set theory is consistent" is then the statement $\neg \psi(a)$ where a is the code for the statement "$0 = 1$." Thus set theory can talk about itself.

Finally we have Gödel's incompleteness theorems for set theory, for whose proofs the reader is referred to any standard logic text:

First Incompleteness Theorem (for Set Theory). There is a formula in the language of set theory which is independent from the axioms of set theory.

Second Incompleteness Theorem (for Set Theory). If set theory is consistent, then it cannot prove its own consistency.

By coding, these are actually statements in set theory itself, e.g., the first incompleteness theorem can be stated within set theory as:

$$\exists x (\phi(x) \wedge \neg \psi(x) \wedge \forall y (\rho(x, y) \rightarrow \neg \psi(y)))$$

where

$\phi(x)$ means x codes a formula

$\psi(x)$ means x codes a theorem

$\rho(x, y)$ means y codes the negation of the formula coded by x.

Many theories are incomplete. For example, the theory of linear orders cannot prove or disprove the statement: $\exists x \forall y \, (x \leq y)$, since some linear orders have a least element and other linear orders do not. For most theories, incompleteness is expected and not interesting. But we make the grandiose claim, which we will actually take some steps to support in chapter 4, that set theory captures our intuition of the mathematical universe; that is, all methods of all branches of mathematics are reducible to set theory. One might hope that every mathematical question we have is thus settled by it, but assuming the consistency of set theory, the first incompleteness theorem says no. There are questions of mathematics that mathematics cannot answer. The second incompleteness theorem adds that one of these undecidable questions is whether set theory itself is consistent.

SECTION 2.3. CARTESIAN PRODUCTS

Before defining ordered n-tuples so that the concept generalizes to infinite dimensions, we need to define relations and functions.

Definition 8

(a) A relation is a set of ordered pairs.[†]

(b) A relation R is a function iff for all x, y, z, $(x, y) \in R$ and $(x, z) \in R$ implies $y = z$.

(c) The domain of a relation R is $\{x : \exists y (x, y) \in R\} = \operatorname{dom} R$.

(d) The range of a relation R is $\{y : \exists x (x, y) \in R\} = \operatorname{range} R$.

(e) The field of a relation $R = \operatorname{field} R = \operatorname{dom} R \cup \operatorname{range} R$.

We say that X is a domain (respectively range or field) of R iff $X \supset \operatorname{dom} R$ (respectively range R or field R).

Note that domains, ranges, fields are not uniquely defined, but *the* domain, range, or field of a relation R is uniquely defined.

For example, if $\operatorname{dom} f = \mathbb{N}$, $f(n) = 2n$ for all n, then range $f =$ the set of even integers, but $\mathbb{R}$ is both a domain and a range of f.

Example 9. Here is the formal definition of equivalence relations: A relation R is an equivalence relation iff

(1) If $x \in \operatorname{field} R$, then $(x, x) \in R$.

(2) If $(x, y) \in R$, then $(y, x) \in R$.

(3) If $(x, y) \in R$ and $(y, z) \in R$, then $(x, z) \in R$.

[†]What we are defining as relations are often called binary relations; what we define as functions are what are called unary functions. For other sorts of relations and functions, see the exercises.

Example 10. Here is the formal definition of partial orders: A relation R is a partial order iff

(1) If $x \in$ field R, then $(x, x) \in R$.
(2) If $(x, y) \in R$ and $(y, x) \in R$, then $x = y$.
(3) If $(x, y) \in R$ and $(y, z) \in R$, then $(x, z) \in R$.

The reader is assumed to be familiar with relations and functions. Let us specify some notation.

Continuation of Definition 8

(f) $f[A] = \{y: \exists x \in A(x, y) \in f\}$ for any relation f.
(g) If $(x, y) \in f$ and f is a function, we write $f(x) = y$.
(h) A relation on a set X is a subset of X^2.
(i) A function f on X is one for which $X = \text{dom } f$.
(j) For any relation R, $R^{-1} = \{(y, x): (x, y) \in R\}$.
(k) A function f is 1–1 iff f^{-1} is a function.
(l) If f is a function, dom $f = A$, and B is a range of f, we write $f: A \to B$.
(m) If $f: A \to B$ and $B = \text{range } f$, we say f is onto B.
(n) If $f: A \to B$ is 1–1 and onto B, we say f is a set isomorphism (or bijection) between A and B.

Using the concepts of definition 8, we now give our official definition of n-tuples and Cartesian products.

Definition 11

(a) An n-tuple $(a_0, \ldots, a_{n-1})$ is a function f with dom $f = \{0, \ldots, n - 1\}$, and $a_i = f(i)$.
(b) The Cartesian product $X_0 \times \cdots \times X_{n-1}$ is the set of all n-tuples f such that $f(i) \in X_i$ for each $0 \le i \le n - 1$.

Notice that an ordered pair (as defined in section 2.2) is not the same as a 2-tuple (as defined in definition 11). Note also that, as we have not defined even 0 yet, we have not officially established that these definitions make sense. Nevertheless, from now on we assume that n-tuples and Cartesian products are defined as in definition 11.

Before generalizing to infinite dimensions we introduce yet more notation: If h is a function with dom $h = I$ and each $h(i) = x_i$ for $i \in I$ we write $\{x_i : i \in I\}$ and use it in two ways:

Strict use: $\{x_i : i \in I\} = h$. That is, we deeply care that x_i is, so to speak, the ith element of the list; x_i's salient feature is that it is $h(i)$. This is sometimes written as $(x_i : i \in I)$ or $\langle x_i : i \in I \rangle$.

Casual use: $\{x_i : i \in I\} = $ range h. That is, we do not particularly care how x_i got into the range of h, although it is convenient to consider it as being the ith element; I is used to index range h.

For example, in casual use $\{2n : n \in \mathbb{N}\}$ is exactly the set of even integers. In strict use $\{2n : n \in \mathbb{N}\}$ is the function $h: \mathbb{N} \to \mathbb{N}$ whose value at n is $2n$. The ambiguity of this notation usually causes no confusion since the use is clear from the context.

Continuation of Definition 11

(c) Given $\{x_i : i \in I\}$ we define the Cartesian product $\prod_{i \in I} x_i = \{f : \text{dom } f = I \text{ and } f(i) \in x_i \text{ for all } i \in I\}$.

Elements of $\prod_{i \in I} x_i$ are called choice functions (because they choose an element out of each x_i). If for some X, each $x_i = X$, we write $\prod_{i \in I} x_i = {}^I X$; note that ${}^I X$ is just $\{f : f \text{ is a function, dom } f = I, \text{ and } X \text{ is a range of } f\}$.

While we will soon establish the existence of ordered n-tuples and the Cartesian product of finitely many sets from elementary axioms, the general existence of choice functions needs its own axiom, the axiom of choice, that will be discussed in chapter 3.

SECTION 2.4. UNION, INTERSECTION, AND SEPARATION

Definition 12

(a) $\bigcup x = \{z : \exists y \in x (z \in y)\}$.

(b) $\bigcap x = \{z : \forall y \in x (z \in y)\}$.

Notice that $x \cup x = \bigcup \{x, x\}$ and is not the same as $\bigcup x$: e.g., let $x = \{y\}$. Then $x \cup x = x$, but $\bigcup x = y$.

When $X = \{x_i : i \in I\}$, we define $\bigcup_{i \in I} x_i = \bigcup X$, $\bigcap_{i \in I} x_i = \bigcap X$.

We are now ready to justify the definitions of section 2.1 by the

Union Axiom. If x is a set, so is $\bigcup x$: $\forall x \exists y (y = \bigcup x)$.

An immediate corollary is that if x, y are sets so is $x \cup y$ (because $\{x, y\}$ is a set, so $\cup\{x, y\}$ is a set). By induction, given any $n \in \mathbb{N}$ and any sets $x_0, \ldots, x_n$, $x_0 \cup \cdots \cup x_n$ is a set.

There is no intersection axiom: Note that $\forall x(x \in \cap\emptyset)$, so if $\cap\emptyset$ were a set, then the collection of all sets would be a set, which we will soon see leads to a contradiction in the presence of other reasonable axioms. To show that $x \cap y$ is a set when x, y are sets we need the

Axiom Schema of Separation. Suppose X is a set and ϕ is a formula. Then $\{x \in X: \phi(x)\}$ is a set. More formally: Let ϕ be a formula in the language of set theory, parameters allowed. Then $\forall x \, \exists y(z \in y$ iff $z \in x$ and $\phi(x))$.

Some authors refer to this as the axiom schema of comprehension. Note that this is a schema, not a single axiom—we need a different axiom for each ϕ.

Let us agree for a moment that "the universe" means "the collection of all sets." Using Russell's paradox we prove

Theorem 13. The universe is not a set: $\neg \exists x \, \forall y(y \in x)$.

Proof. Suppose $\exists x \, \forall y(y \in x)$. Let x be such, and let ϕ be the formula "$z \notin z$." Let $X = \{z \in x: z \notin z\}$. Then $X \in x$; hence $X \in X$ iff $X \notin X$, which is a contradiction.

Theorem 14

(a) $x \neq \emptyset$ iff $\exists y(y = \cap x)$.

(b) If x, y are sets, so are $x - y$ and $x \cap y$.

(c) $x - \cap y = \cup\{x - z: z \in y\}$ if $y \neq \emptyset$.

(d) $x - \cup y = \cap\{x - z: z \in y\}$ if $y \neq \emptyset$.

Parts (c) and (d) are the infinitary De Morgan's laws.

Proof. We do (b): $x - y = \{z \in x: z \notin y\}$; $x \cap y = \{z \in x: z \in y\}$.

We do (c) and leave (d) as an exercise: $w \in x - \cap y$ iff ($w \in x$ and $w \notin \cap y$) iff ($w \in x$ and $\exists z \in y(w \notin z)$) if $\exists z \in y(w \in x$ and $w \notin z)$ iff $w \in \cup\{x - z: z \in y\}$.

Finally, we do (a). Suppose $x = \emptyset$. Then $\forall w(w \notin x)$, so the statement $\forall w(w \in x \rightarrow z \in w)$ is vacuously true for each z. Hence $\forall z(z \in \cap x)$. But this is impossible, so $\cap x$ is not a set.

For the other direction: If $x \neq \emptyset$, then $\forall y \in x(z \in y)$ implies $\exists y \in x(z \in y)$. Hence $\bigcap x = \{z \in \bigcup x : \forall y \in x(z \in y)\}$ which is a set by union and separation.

We close this section with models of union and of separation.

Definition 15. X is a model of union iff $\forall x \in X$ $\exists z \in X(z \cap X = \bigcup\{y \cap X : y \in x \cap X\})$.

For example, let $X = \{a, b, c\}$ where $a = \emptyset$, $b = \{\emptyset\}$, and $c = \{b, \{\{b\}\}\} = \{\{\emptyset\}, \{\{\{\emptyset\}\}\}\}$. We show that X satisfies union. $\bigcup\emptyset = \emptyset$, $\emptyset \cap X = \emptyset$, and $\emptyset \in X$. So $\bigcup\emptyset \in X$. $\bigcup\{\emptyset\} = \emptyset$, so $\bigcup\{\emptyset\} \in X$. Finally, $\bigcup\{\{\emptyset\}, \{\{\{\emptyset\}\}\}\} = \{\emptyset, \{\{\emptyset\}\}\}$ and $\{\emptyset, \{\{\emptyset\}\}\} \cap X = \{\emptyset\} \in X$. So $(\bigcup\{\{\emptyset\}, \{\{\{\emptyset\}\}\}\}) \cap X \in X$, and we are done.

For another example, let $X = \{x_n : n \in \mathbb{N}\}$ where $x_0 = \emptyset$ and each $x_{n+1} = x_n \cup \{x_n\}$. We show that X is a model of union. First note that, by induction, for each $x \in X$, $x \cap X = x$. Then again by induction note that $\bigcup x_0 = \emptyset \in X$ and each $\bigcup x_{n+1} = x_n \in X$. That X is a model of union will then follow from

Proposition 16. X is a model of union if, for all $x \in X$, $x \subset X$ and $\bigcup x \in X$.

Proof. Suppose $x \in X$. Since $x \subset X$, if $y \in x$, then $y \in X$ and $y \subset X$. Let $z = \bigcup x$. Then $z \in X$ and $z \subset X$. So $z \cap X = z = \bigcup\{y : y \in x\} = \bigcup\{y \cap X : y \in x \cap X\}$, as required.

Before discussing models of separation, we need to discuss relativizing a formula ϕ to a set X. The idea is simple: Replace each "$\exists x$" by "$\exists x \in X$," and replace each "$\forall x$" by "$\forall x \in X$." The relativization of ϕ is written ϕ^X. The motivation, as always in models of first-order theories, is that to check whether a formula holds in X you just check the elements of X. Thus, in the language of linear orders, the closed unit interval is a model of the sentence "$\forall x(x \leq 1)$," even though the bigger set $\mathbb{R}$ is not. If ϕ is the formula "$\forall x(x \leq 1)$" and X is the unit interval $[0, 1]$, then ϕ^X is the formula "$\forall x \in [0, 1] (x \leq 1)$."

With this discussion in mind, we are ready for

Definition 17. X is a model of separation iff for all formulas ϕ with parameters in X, $\forall x \in X$ $\exists z \in X$ $(z \cap X = \{y \in x \cap X : \phi^X(y)\})$.

While definition 17 is quite complicated, it has a simple corollary which will suffice for every application we will need.

Proposition 18. X is a model of separation if, for all $x \in X$, if $y \subset x$, then $y \in X$.

The reader is asked to prove proposition 18 as an exercise.

A final, quite technical, note about separation. The alert reader will have noted that the locution "for all formulas ϕ," which occurs in both the definition of the axiom of separation and in definition 17, is not, properly speaking, in the language of set theory. However, using the coding techniques mentioned in section 2.2, we can translate this locution into the language of set theory. Thus, as promised, everything we do can be done within set theory.

SECTION 2.5. FILTERS AND IDEALS

The standard set-theoretic operations enable us to define some combinatorial concepts that will be returned to in chapters 3 and 7. These concepts—of filter and ideal—are of great importance not only in set theory but in topology and measure theory as well; a more generalized definition of ideal is essential in algebra.

Definition 19. A filter on a set X is a family F of subsets of X so that

(a) If $a \in F$ and $X \supset b \supset a$, then $b \in F$. (We say that F is closed under superset.)

(b) If $a_1, \ldots, a_n$ are elements of F, so is $a_1 \cap \cdots \cap a_n$. (We say that F is closed under finite intersections.)

If $\emptyset \notin F$, we say that F is proper. If some $\{x\} \in F$, we say that F is principal; otherwise F is nonprincipal. If, for all $a \subset X$, either a or $X - a$ is an element of F, we say that F is an ultrafilter.

A filterbase on X is a family B of subsets of X whose closure B^s under superset is a filter (i.e., $B^s = \{a \subset X : \exists b \in B(a \supset b)\}$).

Example 20. Let $x \in \mathbb{R}$, and let B be the collection of open intervals $(x - r, x + r)$ where $r > 0$. Then B is a filterbase and B^s is a proper, nonprincipal filter and not an ultrafilter (neither $\{x\}$ nor $\mathbb{R} - \{x\}$ are elements of B^s).

Example 21. Let X be an infinite set, and let $F = \{a \subset X : X - a$ is finite$\}$. Then F is a proper, nonprincipal filter and not an ultrafilter (every infinite set splits into two disjoint infinite subsets).

Example 22. Let X be any nonempty set, choose $x \in X$, and let $F = \{Y \subset X; x \in Y\}$. Then F is a proper, principal ultrafilter on X (since for every $Y \subset X$ either $x \in Y$ or $x \in X - Y$).

If you are wondering where the examples are of proper, nonprincipal ultrafilters, the answer is that such filters need the axiom of choice for their construction. This will be done in chapter 3.

Some facts about ultrafilters:

Lemma 23. Suppose F is an ultrafilter on a set X and $a \in F$. If $b \subset a$, then either $b \in F$ or $a - b \in F$.

Proof. If $b \notin F$, then $X - b \in F$, so $a \cap (X - b) = a - b \in F$.

Lemma 24. Suppose F is a proper ultrafilter on a set X and $X = a_1 \cup \cdots \cup a_n$. Then some $a_i \in F$.

Proof. If not, then each $X - (a_1 \cup \cdots \cup a_k) \in F$, where $k \leq n$, but $X - (a_1 \cup \cdots \cup a_n) = \emptyset$.

Corollary 25. Every proper ultrafilter on a finite set is principal.

Proof. If $X = \{x_1, \ldots, x_n\}$, then $X = \{x_1\} \cup \cdots \cup \{x_n\}$.

The dual concept to a filter is that of an ideal.

Definition 26. An ideal on a set X is a family J of subsets of X so that

   (a) If $b \in J$ and $X \supset b \supset a$, then $a \in J$. (We say that J is closed under subset.)
   (b) If $a_1, \ldots, a_n$ are elements of J, so is $a_1 \cup \cdots \cup a_n$. (We say that J is closed under finite unions.)

The connection between ideals and filters is

Proposition 27. J is an ideal on X iff $F_J = \{X - a : a \in J\}$ is a filter on X.

Proof. That J is closed under finite unions iff F_J is closed under finite intersections follows immediately from De Morgan's laws. That J is closed under subsets iff F_J is closed under supersets follows from the fact that, for a, b subsets of X, $b \supset a$ iff $(X - a) \supset (X - b)$.

We say that J is principal, nonprincipal, or proper according to whether F_J is principal, nonprincipal, or proper. We say J is a maximal ideal iff F_J is an ultrafilter. We say K is a base for an ideal iff K_s, the closure of K under subset, is an ideal.

> **Example 28.** The collection of finite subsets of a given set X is an ideal, proper iff X is infinite. (Example 21 is the filter complementary to this ideal.)

> **Example 29.** Define a nonempty subset Y of $\mathbb{R}$ to be perfect iff it is closed, bounded, and has no isolated points (no isolated points means that every element of Y is the limit of a sequence of distinct points in Y). The collection of perfect subsets of $\mathbb{R}$ is a base for a proper, nonprincipal ideal. Another base for this ideal is the set of bounded closed intervals.

SECTION 2.6. THE NATURAL NUMBERS

We still have not proved that each ordered n-tuple, defined as in 2.11, is a set. Since this definition appeals to the natural number n, we need a set which can represent n in our set-theoretical universe.

Note that we do not claim to settle any philosophical issues, such as "what is zero?" or "in what sense does the number three exist?" Our more modest goal is just to embed our mathematical intuition about natural numbers into the universe of sets so that we can count and do arithmetic within set theory. Later, in chapter 4, we will do the same for $\mathbb{Q}$ and $\mathbb{R}$.

We want the sets we pick to represent natural numbers to be, in some sense, natural choices. To this end we adopt two guiding principles:

Principle A: Each n should have n elements.
Principle B: For each n and m, $n < m$ iff $n \in m$.

These principles, whose serendipity will become clear when we discuss ordinals and cardinals, give us no choice about our definitions.

By principle A, $0 = \emptyset$.

Now suppose we know n and want to define $n + 1$. Since $n < n + 1$, we must have $n \in n + 1$; since $m < n$ implies $m < n + 1$, we must have $n \subset n + 1$. So $n \cup \{n\} \subset n + 1$ by principle B. But by principle A, since n has n elements and $n \cup \{n\}$ has $n + 1$ elements, we cannot put anything more into $n + 1$. So $n + 1 = n \cup \{n\}$.

Thus $1 = \{0\} = \{\emptyset\}$, $2 = \{0, 1\} = \{\emptyset, \{\emptyset\}\}$, $3 = \{0, 1, 2\} = \{\emptyset, \{\emptyset\}, \{0, \{\emptyset\}\}\}$, and so on.

Thus if $\emptyset$ is a set, so is each n.

Now for $\mathbb{N}$. This is a thorny problem. We defined each natural number separately, in terms of its predecessors. This took infinitely many sentences. But we cannot say "$x \in \mathbb{N}$ iff $x = 0$ or $x = 1$ or $x = 2$ or $x = 3$ or ..." because infinite sentences are not allowed. How can we capture our intuition about $\mathbb{N}$ in a finite sentence?

There are many ways to do this. The one we use looks at each natural number as an ordered set. The key properties each natural number x has are:

Property C: Each x is well ordered by $\in$, and if $z \in y \in x$, then $z \in x$ (we say that x is transitive).

Property D: Each nonempty x has exactly one limit element, namely 0.

Property E: Each nonzero x has exactly one element with no successor, which we call $x - 1$, and $x = (x - 1) \cup \{(x - 1)\}$.

A set x satisfying properties C, D, and E will be called a finite ordinal. We define $\mathbb{N}$ to be the collection of finite ordinals. One task of the next section will be to show that $\mathbb{N}$ is a set.

SECTION 2.7. TWO NONCONSTRUCTIVE AXIOMS: INFINITY AND POWER SET

Definition 30. Let x be a set. The successor of x is defined as $x \cup \{x\}$ and is denoted by $S(x)$.

Note that if n is a finite ordinal, $n + 1 = S(n)$.

The Axiom of Infinity. There is a set having zero as an element which is closed under successor: $\exists x (\emptyset \in x$ and $\forall y \in x (S(y) \in x))$.

Notice that this is the first axiom which baldly asserts that sets exist (with the exception of extensionality, our previous axioms had the form "if these are sets, so is this"). An immediate corollary of the infinity axiom is that $\emptyset$ is a set: Let x be as in the axiom of infinity; by separation $\emptyset = \{y \in x: y \neq y\}$. Hence each natural number is a set. Using substitution again, it is easy to show that each n-tuple is a set.

Now we are ready to prove that $\mathbb{N}$ is a set.

Definition 31. A set x is inductive iff $\emptyset \in x$ and $S(y) \in x$ for all $y \in x$.

Theorem 32. Let x be inductive. If n is a finite ordinal, then $n \in x$.

Proof. Suppose there is some finite ordinal $n \notin x$. Then $\{k \in n+1: k \notin x\}$ is nonempty, so it has a least element n^*. Since x is inductive, $0 \in x$, so $n^* \neq 0$. Hence n^* is some $k+1$ where $k \in x$. But since x is inductive, $n^* = k+1 \in x$, contradicting our definition of n^*.

Corollary 33. $\mathbb{N}$ is a set.

Proof. Let x be an inductive set whose existence is guaranteed by the axiom of infinity. Then $\mathbb{N} = \{n \in x: n \text{ is a finite ordinal}\}$.

Set theorists usually refer to $\mathbb{N}$ as ω. We will do this form now on.

The last axiom of this section is the power set axiom.

Definition 34. $\mathscr{P}(x) = \{y: y \subset x\}$ for each x.

Power Set Axiom. If x is a set, so is $\mathscr{P}(x)$: $\forall x \, \exists y (z \in y$ iff $z \subset x)$.

We can now show that each Cartesian product of finitely many sets is a set. For example, $X \times Y = \{(x, y) \in \mathscr{P}(\mathscr{P}(X \cup Y)): x \in X \text{ and } y \in Y\}$.

Let us define models of these axioms.

We want to define models of infinity and power set. So far, when we have defined a model for an axiom, the definition worked for what are known as standard models; that is, what the model thinks is $\in$ really is $\in$. But the definition of standard models of power set and infinity is too complicated for a book on this level. So we will simplify our definition by restricting the class of sets to which it applies, namely, to the class of transitive sets. This is not too great a loss, since every standard model is isomorphic to a transitive model. Transitive sets will be explored extensively in chapter 3. For now, we just mention that x is transitive iff $z \in y \in x$ implies $z \in x$.

Definition 35. Let X be a transitive set.

(a) X is a model of the power set axiom iff $\forall x \in X \exists y \in X (y \cap X = \mathscr{P}(x) \cap X)$.

(b) X is a model of infinity iff $\exists x \in X (\emptyset \in x \cap X \wedge \forall y \in x \cap X (S(y) \in x \cap X))$.

Note that definition 35(b) has an immediate corollary that a transitive set X satisfies infinity if $\omega \in X$ and $\omega \subset X$.

As for 35(a), while it implies that a transitive set X satisfies power set if $\forall x \in X \ \mathscr{P}(x) \in X$, it is far more consequential that this sufficient condition is not necessary. That "$\forall x \in X \mathscr{P}(x) \in X$" may fail, and yet X models power set—that is what makes life interesting.

SECTION 2.8. A DIGRESSION ON THE POWER SET AXIOM

What could be more natural than to gather all the subsets of a given set into a set? But in fact it is this operation that opens the door to most independence results of mathematical interest.

For example, consider $\mathscr{P}(\omega)$. Suppose $\omega \in X$ and X satisfies power set. Then $\mathscr{P}(\omega) \cap X \in X$, but there may be many subsets of ω left out of X. Thus two models of set theory may disagree on how big $\mathscr{P}(\omega)$ is—one thinks it has size ω_1 (to be defined later) and another thinks it has size ω_{27} (which is surely not ω_1). Since there are exactly as many real numbers as subsets of ω (we will prove this later), one model thinks there are exactly ω_1 many reals, another thinks there are exactly ω_{27}, and so on.

Here are four independent statements from four fields of mathematics. You do not have to know what any of the definitions mean to get a sense of the broad sweep of independence results. In all cases, it is the ambiguity about power sets that makes each statement independent.

From general topology: There is a perfectly normal nonmetrizable manifold.

From functional analysis: Every algebraic homomorphism from the Banach space $C[0, 1]$ to an arbitrary Banach space is continuous.

From algebra: Every uncountable Whitehead group is free.

From measure theory: The union of fewer than continuum[†] many measure zero sets has measure zero.

Note that the first two statements appear free from any set-theoretic terminology, and the last two have only the barest trace of set theory (in talking about the size of an object). The last chapter of this book discusses some combinatorial objects which are associated with independent statements.

SECTION 2.9. REPLACEMENT

The final axiom of this chapter—the axiom of replacement, also known as the axiom of substitution—is our last elementary way of introducing sets.

[†]The continuum is **R**. Thus "continuum many sets" means "as many sets as there are reals."

Suppose we have a rule that ought to give a function; i.e., a formula ϕ so that for all x, y, z if $\phi(x, y)$ and $\phi(x, z)$, then $y = z$. (Such a ϕ is called a functional). Some examples of such ϕ are $y = x \cup \{x\}, y = \cup x$, $y = x \cap 17$. Given such a ϕ and a set A we would like there to be a function f where $f = \{(x, y): x \in A$ and $\phi(x, y)\}$. If such an f were to exist, so would its range; range $f = \{y \in \cup \cup f: \exists x \in \cup \cup f((x, y) \in f)\}$.

For historical reasons we go backwards, stating that ranges exist and then proving that the relevant functions also exist.

Replacement Axiom Schema. Ranges of definable functionals exist: If ϕ is a formula so that $\forall x, y, z$ if $\phi(x, y)$ and $\phi(x, z)$, then $y = z$, then $\forall w \exists s(s = \{y: \exists x \in w \, \phi(x, y)\})$.

Note that this is a schema—we have a different axiom for each functional. Note also that ϕ may have parameters, i.e., it may refer to specific sets. (For example, pick any set a. Let ϕ be $y = x \cup a$. Then ϕ qualifies for the replacement schema.)

Theorem 36. Fix ϕ, and suppose $\forall x, y, z \, \phi(x, y)$ and $\phi(x, z)$ implies $y = z$. Fix a set A. Then there is a function f where dom $f = A$ and $f(x) = y$ implies $\phi(x, y)$.

Proof. Given ϕ, A, let s be as in the replacement schema. Then $f = \{(x, y) \in A \times s: \phi(x, y)\}$.

The historical reason for stating the axiom of replacement in terms of images, rather than in terms of functions, is that it was first proposed to show that certain recursive constructions give rise to sets.

Example 37. Define $\omega + \omega = \omega \cup \{S^n(\omega): n \in \omega\}$ where S is the successor operation. Why is $\omega + \omega$ a set? Because of the following functional: $\phi(n, y)$ iff $y = S^n(\omega)$. By replacement and union, we are done.

Example 37 was historically the first use of the axiom of replacement.

Example 38. Define $V_0 = \emptyset$; $V_{n+1} = \mathscr{P}(V_n)$. Now define $V_\omega = \cup_{n \in \omega} V_n$. Why is V_ω a set? Let $\phi(n, y)$ iff $\exists(x_0, \ldots, x_n)(y = x_n, x_0 = \emptyset$, and, for $k < n$, each $x_{k+1} = \mathscr{P}(x_k))$. The reader is invited to prove that ϕ is a functional.

Now let $A = \{y: \exists n \in \omega \, \phi(n, y)\}$. A is a set by replacement, and $V_\omega = \cup A$.

Example 39. Recall the construction of theorem 37 in chapter 1. Here we had a sequence $\{a_n: n \in \omega\}$ of infinite subsets of ω where each $a_n \supset a_{n+1}$, and we constructed a set $A = \{k_n: n \in \omega\}$ to satisfy the property: each $k_n \in a_n$. Now that we have replacement, we can actually prove that A is a set: Let $\psi(x, y)$ iff

  (i) $x = (x_1, \ldots, x_n)$ for some n, $x_1, \ldots, x_n$.
  (ii) y is the least element of $a_{n+1} - \{x_1, \ldots, x_n\}$.
  (iii) Each $x_j \in a_i$ for $j \geq i$.

Now let $\phi(n, y)$ iff $\exists(x_1, \ldots, x_{n-1})[\psi((x_1, \ldots, x_{n-1}), y) \wedge \forall k < n\ \phi(k, x_k)]$. The reader can check that ϕ is a functional and that $\{y: \exists n\ \phi(n, y)\}$ is the desired set A.

The attentive reader will note that the actual construction in chapter 1 was not as restrictive as the construction here—k_n was not required to be the *least* element of a_n not already chosen. The need for restriction (ii) here was to make ϕ a functional; when we have the axiom of choice this sort of restriction—indeed the whole use of replacement in this sort of recursive construction—will become unnecessary.

The final topic in this chapter is models of replacement.

Definition 40. X is a model of replacement iff, for every functional ϕ with parameters in X and every $A \in X$, there is a set $s \in X$ so that $s \cap X = \{y \in X: \exists x \in A \cap X\ \phi^X(x, y)\}$ where ϕ^X is the relativization of ϕ to X as in section 2.4.

The attentive reader may wonder whether ϕ^X is also a functional. The answer is yes, but the proof involves some mathematical logic and will be omitted here.

Corollary 41. Suppose X is transitive. Then X is a model of replacement iff, for every functional ϕ with parameters in X and every $A \in X$, $\{y \in X: \exists x \in A\ \phi^X(x, y)\} \in X$.

For example, we show that V_ω is a model of replacement, using the facts that every element of V_ω is finite and that the image of a finite set is finite.

Suppose $A \in V_\omega$. Then $A \in V_n$ for some n. Now suppose ϕ is a functional with parameters in V_ω. Then $B = \{y \in V_\omega: \phi^{V_\omega}(x, y)$ for some $x \in A\}$ is finite. Let $B = \{y_0, \ldots, y_k\}$ for some k. Then if $i \leq k$, we have n_i with $y_i \in V_{n_i}$. So if $n = \sup\{n_i: i \leq k\}$, $B \subset V_n$, so $B \in V_{n+1}$, hence $B \in V_\omega$.

EXERCISES FOR CHAPTER 2

1. Take the following abbreviated formulas and put them into unabbreviated form:
   (a) $x \cap y \subset x \cup y$.
   (b) $[(x - y) \cup (y - x)] \cap [x \cap y] = \emptyset$.
   (c) $\forall x \exists y (y \notin x)$.
2. (a) Prove that the formulas (a), (b) of exercise 1 hold for all x, y.
   (b) Show that $x \cap y \supset y \cup x$ iff $x = y$.
3. Prove the rest of theorem 3.
4. Let $X = \{E, O\}$ where $E = \{$even integers$\}$ and $O = \{$odd integers$\}$. Show that X does not satisfy extensionality.
5. Let $X = \{x_n: n \in \mathbb{N}\}$ be defined as follows: $x_0 = \emptyset$; $x_{n+1} = x_n \cup \{x_n\}$.
   (a) Show by induction that if $x_n \in x_m \in x_k$ then $x_n \in x_k$.
   (b) Show by induction that if $n < m$ then $x_n \in x_m$.
   (c) Show by induction that if $x_n \in x_m$ then $n < m$.
   (d) Show that $x_n = x_m$ iff $n = m$.
   (e) Show that $x_n \in x_m$ iff $n < m$.
   (f) Show by induction that each x_n has exactly n elements.
   (g) Show that X satisfies extensionality.
   (h) Show that every subset of X satisfies extensionality.
   (*Note*: X is called the set of von Neumann natural numbers, see section 2.6.)
6. Let $X = \{x_n: n \in \mathbb{N}\}$ where $x_0 = \emptyset$ and each $x_{n+1} = \{x_n\}$. Show that if $Y \subset X$ then Y satisfies extensionality iff $Y = \{x_k, x_{k+1}, x_{k+2}, \ldots\}$ for some k or $Y = \{x_k, x_{k+1}, x_{k+2}, \ldots, x_m\}$ for some k, m.

Note: In exercises 7, 9, and 20 you may make use of the following consequence of the axiom of regularity (presented in chapter 3): There are no sets $x, y_0, \ldots, y_n$ with $x \in y_0 \in y_1 \in \cdots \in y_n \in x$.

7. Show that $[x, y] = \{x, \{x, y\}\}$ is reasonable alternative definition of ordered pair; i.e., $[x, y] = [z, w]$ iff $x = z$ and $y = w$.
8. Show that $\emptyset$ satisfies pairing.

9. Show that any model of pairing with at least one element is infinite.

10. This problem deals with a more general definition of relations and a more specific definition of functions than those given in section 2.3. Define $X^1 = X$; define an n-ary relation to be a subset of some X^n and an n-ary function to be a function whose domain is some n-ary relation.

   (a) Show that every set is a unary relation; hence every n-ary relation is also a unary relation.

   (b) Show that every function is a unary function.

   (c) Let R be an n-ary relation where $n \geq 2$. When is $\{((y_1, \ldots, y_{n-1}), y_n): (y_1, \ldots, y_n) \in R\}$ a function?

   (d) In logic it is common to define 0-ary relations as constants. Does this seem reasonable?

11. Recall the definition of a group. There is a set G, a binary operation $\circ$, and an identity element e so that, for all $x, y, z \in G$:

   (a) (Closure under $\circ$) $\exists w \in G(x \circ y = w)$.

   (b) (Associative law) $(x \circ y) \circ z = x \circ (y \circ z)$.

   (c) (e is the identity) $x \circ e = x = e \circ x$.

   (d) (Every element has an inverse) $\exists w(x \circ w = e = w \circ x)$.

   The purpose of this exercise is to translate this into the formal language of set theory. We rewrite $x \circ y$ as $f(x, y)$. We now have a set G, a binary function f, and an element $e \in G$ so that—what? Rewrite axioms (a) through (d) in this new language.

12. Prove the following:

   (a) $\bigcup y \subset x$ if $\forall z \in y(z \subset x)$.

   (b) $x \subset \bigcup y$ iff $\forall z \in x \, \exists w \in y(z \in w)$.

   (c) $\bigcup\{x\} = x$.

13. A set X is closed under union iff $x \in X$ implies $\bigcup x \in X$. Find a set with exactly 7 elements which is closed under union.

14. Show that if X is linearly ordered by $\subset$ then $\forall x, y \in X(x \cup y \in X)$. Find such a set X which is not closed under union; i.e., there is $x \in X$ with $\bigcup x \notin X$.

15. (a) Suppose x is a finite set of natural numbers. Show that $\bigcup x = \max x$.

   (b) Suppose x is an infinite set of natural numbers. Show that $\bigcup x = \omega$.

16. Show that X models separation if $\forall x \in X \, \forall y \subset x(y \in X)$.

17. Show that the family of example 21 is a proper, nonprincipal filter.

18. Show that the family of bounded perfect sets (see example 29) is closed under finite unions and that it is a base for a proper, non-principal ideal.

19. Let $Y = 3 = \{0, 1, 2\}$. Show Y does not model power set.

20. (a) Show that for all x, X, if $x \in X$ then $x \in \mathcal{P}(x) \cap X$.

(b) Show that every nonempty model of power set is infinite. (See the note before exercise 7.)

21. Let x be a set, and let f be defined as follows: $f(0) = x$; $f(n+1) = \bigcup f(n)$ for all $n \in \omega$. Show that both f and its range are sets; hence $\bigcup_{n \in \omega} f(n)$ is a set.

3

REGULARITY AND CHOICE

INTRODUCTION

The axioms of chapter 2 are essentially due to Zermelo (substitution is due to Fraenkel and, independently, to Skolem) and are generally considered the elementary, noncontroversial axioms of set theory. These axioms mostly give ways of constructing sets, of building up the universe, so to speak. In this chapter we discuss the axiom of regularity, which essentially says that these are all the sets there are. The concepts of this axiom developed slowly, starting with Mirimanov in 1917 and ending with von Neumann in 1925. The other axiom discussed in this chapter, the axiom of choice, articulated and defended by Zermelo in 1908, is still controversial—there are mathematicians who do not grant it the same clearly intuitive label that the other axioms have, and when we add it to our list we warn everyone of its presence by using the acronym ZFC. So ZF (= Zermelo–Fraenkel) means all the axioms except choice; ZFC means all of them.

Both regularity and choice cannot really be understood without a reasonable understanding of ordinals. To understand ordinals we must understand transitive sets. So that is our first order of business.

SECTION 3.1. TRANSITIVE SETS

Roughly speaking, a transitive set keeps no secrets from itself. It knows all there is to know about its elements; what you see is what you get.

Definition 1. X is transitive iff $\forall y \in X(y \subset X)$.

In other words, if X is transitive and $z \in y \in X$, then $z \in X$. Note that every natural number is transitive.

Lemma 2. Every transitive set models extensionality.

Proof. Suppose X is transitive. If $x, y \in X$ and $x \neq y$, then there is some $a \in (x - y) \cup (y - x)$. By transitivity, $a \in X$, and we are done.

Lemma 2 says that if a transitive set thinks two sets have the same elements then in fact they do. Similarly, if a transitive set thinks another set is empty, then indeed it is, and so on, for pairs, ordered pairs, unions, relations, functions, etc.

Lemma 3. Let X be transitive.

    (a) If $y \in X$ and $y \cap X = \emptyset$, then $y = \emptyset$.
    (b) If $a, x, y \in X$ and $a \cap X = \{x, y\}$, then $a = \{x, y\}$.
    (c) If $a, x, y \in X$ and $a \cap X = (x, y)$, then $a = (x, y)$.
    (d) If $a, x \in X$ and $a \cap X = \bigcup x$, then $a = \bigcup x$.
    (e) If $a \in X$ and $a \cap X$ is a relation, then a is a relation.
    (f) If $a \in X$ and $a \cap X$ is a function, then a is a function.

Proof. All parts of lemma 3 are corollaries of the following fact about transitive sets: If X is transitive and $a \in X$, then $a \cap X = a$.

Let us try to build a transitive set. $\emptyset$ is of course transitive, as is $\{\emptyset\}$. Suppose we want a slightly more complicated transitive set, one, e.g., with $\{\emptyset\}$ as an element. Then $\emptyset$ must also be an element, and that suffices: $\{\emptyset, \{\emptyset\}\}$ is transitive. In general, if x is a transitive set, then so is $S(x)$ (see the exercises).

Lemma 4. X is transitive iff $\bigcup x \subset X$ for all $x \subset X$.

Proof. Suppose X is transitive, $x \subset X$. If $y \in x$, then $y \in X$, so if $z \in y \in x$, then $z \in X$. Hence $\bigcup x \subset X$. On the other hand, if for all $x \subset X$, $\bigcup x \subset X$, and if $y \in X$, then $\{y\} \subset X$, so $\bigcup \{y\} \subset X$, but $\bigcup \{y\} = y$, so $y \subset X$ and X is transitive.

Lemma 4 gives us a standard way to construct transitive sets.

Definition 5. The transitive closure of a set x (written $TC(x)$) is $(x \cup \bigcup x \cup \bigcup \bigcup x \cup \cdots)$.

More formally, define f_x by $f_x(0) = x$; $f_x(n+1) = \bigcup f_x(n)$ for all $n \in \omega$. Then $TC(x) = \bigcup_{n \in \omega} f_x(n)$.

Theorem 6. Let x be a set. Then $TC(x)$ is transitive, and if $x \in y$ and y is transitive, then $TC(x) \subset y$.

Proof. Suppose $z \in TC(x)$, $w \in z$. Then $z \in f_x(n)$ for some n, so $w \in f_x(n+1)$, hence $w \in TC(x)$. Thus $TC(x)$ is transitive.

If $x \in y$ and y is transitive, then $f_x(0) = x \subset y$, and, by the transitivity of y, if $f_x(n) \subset y$, then $f_x(n+1) \subset y$. So by induction each $f_x(n) \subset y$, hence $TC(x) \subset y$.

Corollary 7. A set x is transitive iff $x = TC(x)$.

Proof. $TC(x)$ is transitive, and $x \subset TC(x)$ for any x, so all we have to prove is that if x is transitive then $TC(x) \subset x$. By transitivity of x, each $f_x(n) \subset x$. By induction, $TC(x) \subset x$.

SECTION 3.2. A FIRST LOOK AT ORDINALS

Recall from chapter 1 the strict order $<$ derived from a partial order $\le$. We say that a set X is strictly well-ordered by a relation $E \subset X^2$ if the relation E^+ given by

$$x E^+ y \quad \text{iff} \quad x E y \quad \text{or} \quad x = y$$

is a well-ordering and E is the resulting strict order; i.e., if $x E y$, then $x E^+ y$ and $x \ne y$.

Definition 8. An ordinal is a transitive set strictly well-ordered by $\in$.

Note that each finite ordinal is an ordinal.
We usually reserve Greek letters for ordinals. We will shortly prove that ω is an ordinal; this justifies our renaming of $\mathbb{N}$.

Definition 9. Let α, β be ordinals. We write $\alpha < \beta$ iff $\alpha \in \beta$.

For example, $S(\omega) = \omega \cup \{\omega\}$ is an ordinal, called (not surprisingly) $\omega + 1$. In general, if α is an ordinal, $S(\alpha)$ is written as $\alpha + 1$, and we write $S(\alpha + n)$ as $\alpha + n + 1$. Clearly each $\alpha < \alpha + 1$, and there is no ordinal β with $\alpha < \beta < \alpha + 1$.

Lemma 10. If α is an ordinal and $\beta \in \alpha$, then β is an ordinal.

Proof. Suppose $\beta \in \alpha$. If $\beta - \alpha \neq \emptyset$, then α is not transitive, so $\beta \subset \alpha$, and β is strictly well-ordered by $\in$. Suppose $\gamma \in \beta$ and $\delta \in \gamma$. Since α is linearly ordered by $\in$, by the transitivity of linear orderings, $\delta \in \beta$. So β is transitive.

Lemma 11. If α is an ordinal, then $\alpha \notin \alpha$.

Proof. Otherwise $\alpha \in \alpha$; i.e., $\in$ would not be a strict order.

Definition 12. A is an initial segment of an ordered set L iff $A \subset L$ and for all $t \in A$ if $s < t$ then $s \in A$.

For example, the open interval $(-\infty, a)$ is an initial segment of $\mathbb{R}$, for all $a \in \mathbb{R}$.

Lemma 13. If A is an initial segment of an ordinal α, then either $A \in \alpha$ or $A = \alpha$.

Proof. That A is an ordinal follows by a proof similar to that of lemma 10. Notice that for all $\beta \in \alpha$ either $\beta \in A$ or $\beta > \gamma$ for all $\gamma \in A$. Suppose $A \neq \alpha$. Then there is $\beta \in \alpha$ with $A \subset \beta$. Pick the least such β and suppose $A \neq \beta$. Then there is $\delta \in \beta$ with $A \subset \delta$. But by transitivity of α, $\delta \in \alpha$, so β was not minimal, which is a contradiction. Hence $A = \beta$, and so $A \in \alpha$.

Theorem 14. If α, β are ordinals, then $\alpha < \beta$, $\alpha = \beta$, or $\beta < \alpha$.

Proof. Given distinct α, β let $A = \alpha \cap \beta$. A is an initial segment of α, β by transitivity, so A is an ordinal. Either $A \in \alpha$ or $A = \alpha$. Either $A \in \beta$ or $A = \beta$. If $A \in \alpha$ and $A \in \beta$, then $A \in A$, contradicting lemma 11. So either $A = \alpha \in \beta$ or $A = \beta \in \alpha$ or $A = \alpha = \beta$.

Theorem 15. ω is an ordinal.

Proof. By definition, ω is a set of ordinals, so it is strictly well-ordered by $\in$. We must show that ω is transitive. Suppose it is not. Let $A = \{\alpha \in \omega : \alpha \not\subset \omega\}$. If A is nonempty, then it has a least element α^*. Then $\alpha^* \neq \emptyset$. Suppose $\alpha^* = \beta + 1$ for some $\beta \in \omega$. Then $\beta \subset \omega$ and $\alpha^* = \beta \cup \{\beta\} \subset \omega$, a contradiction. So $\omega - \{\alpha^*\}$ is inductive. But when we proved that each inductive set had every finite ordinal as an element we were proving that $\omega \subset x$ for each inductive set x. Hence $\omega \subset \omega - \{\alpha^*\}$, which is a contradiction.

Thus the first few ordinals are $0, 1, 2, 3, \ldots \omega, \omega + 1, \omega + 2, \omega + 3, \ldots$.

All of chapter 5, most of chapter 6, and much of chapter 7 are concerned with ordinals, including quasi-concrete examples. For now we content ourselves with a few more useful facts.

Theorem 16 (Induction on Ordinals)

Version I (*Set Version*). Let α be an ordinal, ϕ, a formula, and suppose we know that for all $\beta \in \alpha$ if $\phi(\gamma)$ holds for all $\gamma < \beta$ then $\phi(\beta)$. Then for all $\beta \in \alpha$, $\phi(\beta)$.

Version II (*Class Version*). Let ϕ be a formula, and suppose we know that for all ordinals β if $\phi(y)$ holds for all $\gamma < \beta$ then $\phi(\beta)$ holds. Then $\phi(\beta)$ holds for every ordinal β.

Proof. Suppose not. Then there is a counterexample γ so that $\phi(\gamma)$ fails. (For version I we insist that $\gamma < \alpha$.) Let δ be the least counterexample; i.e., δ is the least ordinal in $\{\beta \leq \gamma : \neg\phi(\beta)\}$. Then $\phi(\beta)$ holds for all $\beta < \delta$. Hence by hypothesis $\phi(\delta)$ holds, which is a contradiction.

Notes on Induction

1. The hypothesis on ϕ implies $\phi(0)$ (since "$\forall \beta < 0 \; \phi(\beta)$" is vacuously true).
2. If $\alpha = \omega$, then version I is equivalent to the usual induction on $\mathbb{N}$.

Closely related to induction is recursive construction on the ordinals. This generalizes recursive constructions on $\mathbb{N}$ just as inductive proofs generalize induction on $\mathbb{N}$. As with induction, there are both set versions and class versions of recursive constructions on ordinals. Here is how a set recursive construction works. Given some ordinal α, a collection of sets $\{X_\beta : \beta < \alpha\}$ is described via some functional ϕ. By substitution this collection is a set; hence so is $X = \bigcup \{X_\beta : \beta < \alpha\}$. Theorem 16 is then applied to the specific situation to show that, since each X_β satisfies a given induction hypothesis, X has the desired properties. Class recursive construction simply uses set recursion at each stage to eventually move through all the ordinals. Examples of recursive construction can be found in our uses of the axiom of choice, in the definition of the V_α's in section 3.3, and in later chapters.

Some notation: We use ON to denote the collection of all ordinals. "ON" is an abbreviation, and we must be careful how we use it since

Theorem 17. ON is not a set.

Proof. Suppose $\exists x(x = \text{ON})$. Then x is transitive and well-ordered by $\in$, so x is an ordinal, so $x \in x$, which is a contradiction.

Legal uses of ON are in phrases like $x \in \bigcup_{a \in \text{ON}} \mathcal{P}(\alpha)$, which is shorthand for $\exists \alpha$ (α is an ordinal and $x \subset \alpha$). Illegal uses are in phrases like $\text{ON} \in x$.

Note that every set of ordinals has a least element. Thus, by theorem 14, we can, loosely speaking, say that ON is well-ordered.

Corollary 18. Every set of ordinals is a subset of some ordinal; i.e., $\forall x \, \exists \alpha \in \text{ON}(x \cap \text{ON} \subset \alpha)$.

The proof of corollary 18 uses the following useful observation:

Corollary 19. If x is a set of ordinals, then $\bigcup x$ is an ordinal.

Proof of Corollary **19.** Since $\bigcup x$ is a subset of ordinals, it is strictly well-ordered by $\in$. If $\delta \in \gamma \in \bigcup x$, then $\gamma \in \beta$ for some $\beta \in x$, so $\delta \in \beta$, so $\delta \in \bigcup x$, hence $\bigcup x$ is transitive.

If $x \subset \text{ON}$, we define $\sup x = \bigcup x$.
Note that $\sup x =$ the least ordinal α with $\beta \leq \alpha$ for all $\beta \in x$.

Proof of Corollary **18.** Let x be a set and let $y = x \cap \text{ON}$. By separation, y is a set, and since y is a set of ordinals, $\bigcup y$ is an ordinal, say α. But then if $\beta > \alpha$, $\beta \notin y$, so $y \subset \alpha + 1$, and we are done.

SECTION 3.3. REGULARITY

Axiom of Regularity. Every nonempty set has a minimal element (with respect to $\in$): $\forall x \neq \emptyset \, \exists y \in x(x \cap y = \emptyset)$.

The axiom of regularity has two general sorts of consequences, local and global. Let us look at the local consequences first.

Theorem 20. $\forall x(x \notin x)$.

Proof. If $x \in x$, then $\{x\}$ has no minimal element.

Theorem 21. There is no infinite descending $\in$-chain; i.e., if $\{x_n : n \in \omega\}$ is a sequence of sets, then $\forall n(x_{n+1} \in x_n)$ is not possible.

Proof. Otherwise $\{x_n : n \in \omega\}$ has no minimal element.

In fact, the statement of theorem 21 is equivalent to the axiom of regularity:

Theorem 22. There is no infinite descending $\in$-chain iff every nonempty set has a minimal element (with respect to $\in$).

Note. The point of theorem 22 is that the axiom of regularity is equivalent to the statement "there is no infinite descending $\in$-chain," so we cannot use regularity in the proof of sufficiency.

Proof. One direction is just theorem 21. For the other direction, suppose regularity fails. Let x be a set with no minimal element. Let $x_0 \in x$. Then x_0 is not minimal, so there is $x_1 \in x_0 \cap x$. And x_1 is not minimal, so there is $x_2 \in x_1 \cap x \ldots$, and so on. Thus we have $\ldots x_2 \in x_1 \in x_0 \in x$, i.e., an infinite descending $\in$-chain.

There is only one problem with this proof: We have used a subtle form of the axiom of choice. Given x we have many choices of potential x_0's. Given each x_0 we have many choices of potential x_1's. And so on. How can we pick a path through this maze? I think we are all agreed that we can, but the ability to do it relies on yet another axiom, a weak form of the axiom of choice called dependent choice. Since the system we will ultimately use is full ZFC, we can let theorem 23 stand as is, but we must recognize that we have invoked a new axiom to do it. In fact, it can be shown that some form of the axiom of choice is needed to prove theorem 22.

The global consequences of regularity need more definitions.

In keeping with definition 30 in chapter 1 we have

Definition 23. An ordinal α is a limit ordinal if $\forall \beta \in \text{ON} \; \alpha \neq \beta + 1$. If α is not a limit, it is called a successor.

Theorem 34 in chapter 1 gives us our first picture of what ON looks like—a very very very long string of copies of ω where the first element in each copy is a limit ordinal and the rest are successors.

Definition 24. Define $V_0 = \emptyset$. If α is an ordinal, then $V_{\alpha+1} = \mathscr{P}(V_\alpha)$. If α is a limit ordinal, then $V_\alpha = \bigcup_{\beta < \alpha} V_\beta$.

Definition 24 is a good example of a class recursive construction, generalizing our definition of V_ω in chapter 1.

Lemma 25. Each V_α is transitive, and if $\alpha < \beta$, then $V_\alpha \subset V_\beta$.

The proof of lemma 25 is left as an exercise. We use the lemma, however, in

Lemma 26. If $\alpha < \beta$, then $V_\alpha \in V_\beta$.

Proof. By induction on β. Suppose it is true for all $\gamma < \beta$; i.e., if $\gamma < \beta$ and $\alpha < \gamma$, then $V_\alpha \in V_\gamma$. If β is a limit ordinal, then $V_\beta = \bigcup_{\gamma < \beta} V_\gamma$. Hence by induction hypothesis if $\gamma < \beta$, then $V_\gamma \in V_{\gamma+1} \subset V_\beta$. Otherwise $\beta = \gamma + 1$ for some γ and $V_\beta = \mathscr{P}(V_\gamma)$. Let $\alpha < \beta$. If $\alpha = \gamma$, then $V_\gamma \in \mathscr{P}(V_\gamma)$, so $V_\gamma \in V_\beta$. If $\alpha < \gamma$, then $V_\alpha \in V_\gamma$ by induction hypothesis, so $V_\alpha \subset V_\gamma$ by transitivity, so $V_\alpha \in V_\beta$.

The next theorem says that the V_α's are the way to build up the universe.

Theorem 27. $\forall x \, \exists \alpha \in \mathrm{ON}(x \in V_\alpha)$.

Note. We write this as $V = \bigcup_{\alpha \in \mathrm{ON}} V_\alpha$, where V is understood to abbreviate "the universe of sets."

Proof. Suppose $a \notin \bigcup_{\alpha \in \mathrm{ON}} V_\alpha$, that is, $\forall \alpha \in \mathrm{ON}(a \notin V_\alpha)$. Suppose $a \subset \bigcup_{\alpha \in \mathrm{ON}} V_\alpha$. Then for each $x \in a$ we assign the function rank x by rank x = the least α such that $x \subset V_\alpha$. The range of the rank function is a set of ordinals, call it R, and R is a subset of some ordinal β by corollary 18. Hence $a \subset V_\beta$, so by definition 24, $a \in V_{\beta+1}$, which is a contradiction. So we assume $a \not\subset \bigcup_{\alpha \in \mathrm{ON}} V_\alpha$.

By transitivity of the V_α's, $\mathrm{TC}(a) \not\subset \bigcup_{\alpha \in \mathrm{ON}} V_\alpha$. Without loss of generality, assume a is transitive.

There is some $x \in a$, $x \notin \bigcup_{\alpha \in \mathrm{ON}} V_\alpha$. Let $b = \{x \in a : x$ transitive and $x \notin \bigcup_{\alpha \in \mathrm{ON}} V_\alpha\}$. Since $b \neq \emptyset$, it has a minimal element y. Since $y \notin \bigcup_{\alpha \in \mathrm{ON}} V_\alpha$, we can argue again that $y \not\subset \bigcup_{\alpha \in \mathrm{ON}} V_\alpha$, so there is $z \in y - \bigcup_{\alpha \in \mathrm{ON}} V_\alpha$. But then $z \in y \cap b$, so y was not minimal which is a contradiction.

Let us explore the structure of the V_α's.

Lemma 28. Each V_α is a set.

Proof. The V_α's are defined via power set, substitution, and union.

Definition 29. For each $x \in V$ define rank x = the least α with $x \subset V_\alpha$.

Note that if rank $x = \alpha$ then $\beta = \alpha + 1$ is the least ordinal for which $x \in V_\beta$. In particular, if $x \in V_\beta$, then rank $x < \beta$.

Lemma 30. If $x \in y$, then rank $x <$ rank y.

Proof. Suppose rank $y = \beta$. Then $y \notin V_\beta$ and $y \subset V_\beta$. So $x \in V_\beta$, hence rank $x < \beta =$ rank y.

An easy corollary (see the exercises) is that each rank $\alpha = \alpha$.

Suspending the axiom of regularity for a moment, we use the V_α's to characterize it.

Theorem 31. $V = \bigcup_{\alpha \in \text{ON}} V_\alpha$ iff every nonempty set has a minimal element (with respect to $\in$).

Proof. Sufficiency is just theorem 27. To prove necessity, of course, we may not use regularity, since that is what we are trying to prove.

So suppose $V = \bigcup_{\alpha \in \text{ON}} V_\alpha$, and let $x \in V$. (Recall: "$x \in V$" is just shorthand for "x is a set.") We may assume $x \neq \emptyset$. Let a be the range of the rank function on x; i.e., $a = \{\alpha : \exists y \in x \text{ rank } y = \alpha\}$. Let α be minimal in a. Let $y \in x$, rank $y = \alpha$. Then $y \cap x = \emptyset$, since if $z \in y \cap x$, rank $z <$ rank y, which contradicts the minimality of α.

Thus the axiom of regularity is equivalent to the statement "$V = \bigcup_{\alpha \in \text{ON}} V_\alpha$." It gives a clear picture of how the universe is built up. We will assume the axiom of regularity from now on.

Example 32. $V_1 = \{\emptyset\}$. $V_2 = \{\emptyset, \{\emptyset\}\} = 2$. $V_3 = \{\emptyset, \{\emptyset\}, \{\emptyset, \{\emptyset\}\}, \{\{\emptyset\}\}\}$. You may find V_4 as an exercise. As this exercise shows, the V_n's get large very quickly.

Example 33. Suppose rank $x =$ rank $y = \alpha$. Then rank$(\{x, y\}) = \alpha + 1$, rank$((x, y)) = \alpha + 2$, rank$(x \cup y) = \alpha$, and rank$(\bigcup x) \leq \alpha$.

Let us prove just the last one: rank$(\bigcup x) \leq \alpha$. Since rank $x = \alpha$, $x \subset V_\alpha$. If $w \in z \in x$, then rank $z < \alpha$, so rank $w <$ rank $z < \alpha$, so $w \in V_\alpha$, so $\bigcup x \subset V_\alpha$, and we are done.

We will return to the V_α's later, to check them as potential models of set theory.

SECTION 3.4. A WORD ABOUT CLASSES

Our uses of ON and V are examples of class notation. That is, we have collections which are too big to be called sets but are so easy to describe that we find ourselves using shorthand when referring to them. That is what we shall mean by a class.

Informal Definition 34. Let ϕ be a formula. The class defined by ϕ is the collection of all x such that $\phi(x)$ holds.

Here are some more examples of classes: $\{x: \text{rank } x > \omega\}$; $\{\alpha \in \text{ON}: \alpha$ is a limit$\}$; $\{x: x \text{ is transitive}\}$; $\{V_\alpha: \alpha \in \text{ON}\}$.

Classes are not elements of our universe. Every time one pops up in a formula it can be gotten rid of by using the formula which defines it.

There are versions of set theory which do incorporate classes as an integral notion, the most popular one being Gödel–Bernays set theory (GB). But the ones which are commonly accepted can be shown to prove no more about sets than ZF does, and they carry a more cumbersome logical baggage. We lose nothing by restricting ourselves to ZF.

SECTION 3.5. THE AXIOM OF CHOICE

Axiom of Choice (AC). A product of nonempty sets is nonempty: If $I \neq \emptyset$ and, given the family $\{x_i: i \in I\}$, each $x_i \neq \emptyset$, then $\prod_{i \in I} x_i \neq \emptyset$. An element of $\prod_{i \in I} x_i$ is called a choice function.

To give a more tangible image: If we think of each x_i as a pot, then we may dip down simultaneously into each pot and pull something out of each.

Let us see where AC is not needed: If $X \neq \emptyset$, then, for any $I \neq \emptyset$, we already know that $^IX \neq \emptyset$: pick $a \in X$ and let $f(i) = a$ for all i. The axiom of choice is not needed because we pick just once from only one pot.

The axiom of choice is not needed either if I is finite: We have already shown that the Cartesian product of finitely many nonempty sets is nonempty.

Nor is the axiom of choice needed when the sets carry a strong enough structure. For example, if each X_i is well-ordered, we can define $f(i)$ as the least element of X_i. By substitution, f is a function. Similarly, if each X_i is a ring with unity e_i, we can define $f(i) = e_i$ for all i. And so on.

If we have infinitely many pairs of shoes, we do not need the axiom of choice to pick one shoe from each pair: Just pick the left shoe. But if we have infinitely many pairs of socks and want to pick one sock from each pair—then we need the axiom of choice.

AC is embedded deeply in standard mathematics. We use choice to prove things that "ought" to be true, e.g., theorem 22 in this chapter, theorem 29 in chapter 1, the existence of a completion for each field, the existence of a basis for each vector space. It seems so natural that we often do not recognize we have used it.

I will admit my bias here—the axiom of choice seems obviously true to me. If you share my bias, you need to be told why the absence of choice is worth thinking about.

The first two reasons are of fairly recent origin. It is possible to prove theorems about ZFC by starting with models of ZF in which strong combinatorial principles hold which negate choice and then moving to models in which choice holds. Further, given strong enough assumptions about large cardinals (we will meet some large cardinals in chapter 5), there are natural models of ZF in which choice fails spectacularly.

The third reason goes back to the origins of axiomatic set theory: Some people just do not like being cavalier about infinite processes. "Choose?" they say, "*how* do you choose? What rule do you use?" And they point to equivalent forms of the axiom of choice which are even more transparently nonconstructive. The chief of these is

The Well-Ordering Principle (WO). Every set can be well-ordered: $\forall x \exists \alpha \in \text{ON} \exists f(f$ is a 1–1 function from α onto $x)$.

"Show me a well-ordering of $\mathbb{R}$!" these people cry. And as it turns out, without some form of AC you cannot.

AC holds the same relation to ZF as the parallel postulate holds to the other Euclidean axioms. It seems unnecessary, as if it ought to follow from the other axioms. It bothers people. And for a long time (although it took much less than thousands of years) people tried to show that either it followed from or was negated by ZF.

In 1935 Gödel showed that ZFC is consistent if ZF is. In 1963 Cohen showed that $ZF + \neg AC$ is consistent if ZF is. So AC is independent from ZF.

In the next section we prove that several statements are equivalent to the axiom of choice. After that we do not really need it until chapter 5, when we discuss cardinals. We will try to do without it there as long as we can, just to develop our sense of what can be done without it, but will quickly give up (very little in that chapter can be done without it) and from then on will simply assume ZFC for whatever we do.

SECTION 3.6. FOUR FORMS OF THE AXIOM OF CHOICE

We have already stated AC and WO. We will state two more forms of
the axiom of choice, give applications, and then prove our forms
equivalent.

Zorn's Lemma (ZL). If every nonempty chain in a nonempty partial
order P has an upper bound, then P has a maximal element.

Hausdorff's Maximal Principle (HMP). Every nonempty chain in a
partially ordered set can be extended to a maximal chain.

Here a maximal chain is a chain C so that if $x \notin C$ then $\{x\} \cup C$ is not
a chain. Equivently, C is a maximal element of the set of chains under $\subseteq$.
Let us give two applications of WO, ZL, and HMP.

Definition 35. If X is linearly ordered by $\leq$, we say A is cofinal in X iff
$A \subseteq X$ and for every $x \in X$ there is $y \in A$, $y \geq x$.

Some examples: (1) $\mathbb{N}$ is cofinal in $\mathbb{Q}$ which is cofinal in $\mathbb{R}$. (2) Let
X be linearly ordered with no endpoint. Then X is cofinal in $L(X)$,
the lexicographic order on $\bigcup_{n<\omega} X^n$. (3) If X has a maximal
element m, then A is cofinal in X iff $m \in A$.

Theorem 36. Every linear order has a well-ordered cofinal subset.

Proof

Method I (WO). Given a linearly ordered set X, by WO we may write
$X = \{x_\alpha : \alpha < \beta\}$ for some ordinal β. (Note that the order in which we list
the elements of X has nothing to do with the original order on X.) The
desired well-ordered cofinal subset of X is $A = \{x_\gamma : \forall \delta < \gamma(x_\delta < x_\gamma)\}$.
Note that $x_0 \in A$, so A is nonempty. We must show that A is cofinal and
well-ordered.

If $x = x_\alpha \in X$, then either $\exists \gamma < \alpha$ with $x_\gamma > x_\alpha$ or $x_\alpha \in A$. If $x_\alpha \notin A$, we
let γ be the least $\gamma < \alpha$ with $x_\gamma > x_\alpha$. Then if $\delta < \gamma$, $x_\delta < x_\gamma$, so $x_\gamma \in A$,
and we have proven A cofinal.

By definition, if $x_0 < x_\alpha < x_\gamma$ and $x_\alpha, x_\gamma \in A$, then $\alpha < \gamma$. Hence a
descending chain in A corresponds to a descending chain in ON, so A is
well-ordered.

Method II (ZL). Let $\mathcal{A}$ be the collection of all well-ordered subsets of X

ordered by end-extension: $A \leq B$ if A is an initial segment of B. Note that $\mathcal{A}$ has no infinite descending chains, since if $A_1 > A_2 > A_3 \ldots$ and $x_i \in A_i - A_{i+1}$ then $x_1 > x_2 > \cdots$, hence A_1 would not be well-ordered. A chain $\mathcal{C}$ in $\mathcal{A}$ is a collection of well-ordered subsets so that if $A, B \in \mathcal{C}$ then either $A \leq B$ or $B \leq A$. To apply ZL we must show each nonempty chain has an upper bound. Let $\mathcal{C}$ be a nonempty chain. Then $\bigcup \mathcal{C} \geq A$ for each $A \in \mathcal{C}$, $\bigcup \mathcal{C}$ is linearly ordered, and if $\bigcup \mathcal{C}$ had an infinite descending chain $x_1 > x_2 > \cdots$, then if $A \in \mathcal{C}$ where $x_1 \in A$, A would have an infinite descending chain, a contradiction. So $\bigcup \mathcal{C}$ is well-ordered, hence $\bigcup \mathcal{C} \in \mathcal{A}$, and $\bigcup \mathcal{C}$ is an upper bound for $\mathcal{C}$. By ZL $\mathcal{A}$ has a maximal element A. By maximality of A, if $x \in X$, there is $y \in A$ with $y \geq x$. So A is as desired.

Method III (HMP). Let $\mathcal{A}$ be as in method II, and by HMP let $\mathcal{C}$ be a maximal chain in $\mathcal{A}$. We already know that $\bigcup \mathcal{C}$ is a well-ordered subset of X; since $\mathcal{C}$ is maximal, $\bigcup \mathcal{C}$ is cofinal.

Notes on Methodology

1. Method I (WO) has a slightly constructive flavor. Set theorists often prefer proofs from WO because they make the inductive process clear. Using ZL or HMP is slicker, but it obscures what is going on.

2. In using ZL the sticking point is in showing ZL actually applies; i.e., every nonempty chain has an upper bound. Be careful.

3. The similarity of methods II and III, just as the similarity in the statements ZL and HMP, should not have gone unnoticed by the reader.

4. The reader will have noticed that we have given no direct proof from AC. This is typical. Only when justifying an argument such as that used to prove theorem 22 is AC quoted directly. Its chief use is to look intuitive so that we will accept it and hence its disguises.

To give us more practice, here is another theorem which depends on AC, again proved in three different ways. Recall the discussion of filters in section 2.5. As promised, we prove

Theorem 37. If F is a proper filter on a set x, then it extends to an ultrafilter on x; i.e., there is an ultrafilter G on x, $G \supset F$.

Proof

Method I (WO). Let F be a proper filter on x. By WO we enumerate $\mathcal{P}(x)$ as $\{y_\alpha : \alpha < \beta\}$ for some ordinal β.

We will construct a sequence $\{F_\alpha: \alpha < \beta\}$ of subsets of $\mathscr{P}(x)$ so:

(a) $F_0 = F$.
(b) If $\alpha < \delta$, then $F_\alpha \subset F_\delta$.
(c) If A is a finite subset of some F_α, then $\bigcap A \in F_\alpha$, $\bigcap A \neq \emptyset$.
(d) If $\alpha < \beta$, then either $y_\alpha \in F_{\alpha+1}$ or $x - y_\alpha \in F_{\alpha+1}$.

You can check that these properties assure that $\bigcup_{\alpha < \beta} F_\alpha$ is an ultrafilter.

We know what F_0 is. Suppose we know F_γ for all $\gamma < \alpha$. If α is a limit ordinal, let $F_\alpha = \bigcup_{\gamma < \alpha} F_\gamma$. (You should check that F_α has the required properties.) Suppose $\alpha = \gamma + 1$ for some γ. If $y_\gamma \cap a \neq \emptyset$ for all $a \in F$, let $F_\alpha = F_\gamma \cup \{y_\gamma\} \cup \{y_\gamma \cap a: a \in F_\gamma\}$. It is easy to check (a), (b), (d); (c) follows because finite intersections from F_γ are in F_γ, so finite intersections from $F_{\gamma+1}$ have the form $y_\gamma \cap a$ where $a \in F_\gamma$. On the other hand if $y_\gamma \cap a = \emptyset$ for some $a \in F_\gamma$, then $x - y_\gamma \supset a$, so $(x - y_\gamma) \cap b \supset a \cap b$ for all $b \in F_\gamma$, hence $(x - y_\gamma) \cap b \neq \emptyset$ for all $b \in F_\gamma$. Let $F_{\gamma+1} = F_\gamma \cup \{x - y_\gamma\} \cup \{(x - y_\gamma) \cap b: b \in F_\gamma\}$. Again, (a) through (d) are satisfied.

Method II(ZL). Let $\mathscr{F} = \{G: G$ is a proper filter on $x, F \subset G\}$. Partially order $\mathscr{F}$ by $\subset$. If $\mathscr{C}$ is a nonempty chain in $\mathscr{F}$, $\bigcup \mathscr{C} \in \mathscr{F}$; otherwise you would have $a_1, \ldots, a_n, G_1, \ldots, G_n, a_i \in G_i \in \mathscr{C}, G_i \subseteq G_{i+1}$ for $i < n$, $\bigcap_{i<n} a_i \notin G_n$, which contradicts G_n being a proper filter. So the hypothesis of Zorn's lemma is met, and $\mathscr{F}$ has a maximal element G. We must show that G is an ultrafilter.

Suppose $y \notin G$ for some $y \subset x$. By the maximality of G there is $a \in G$ with $y \cap a = \emptyset$. So $x - y \supset a$, hence $x - y \in G$.

Method III(HMP). Let $\mathscr{F}$ be as in method II, and let $\mathscr{C}$ be a maximal chain in $\mathscr{F}$. By the methods of method II, $\bigcup \mathscr{C} \in \mathscr{F}$; by the maximality of $\mathscr{C}$, $\bigcup \mathscr{C}$ is an ultrafilter.

Finally, we have

Theorem 38. ZL $\leftrightarrow$ HMP $\leftrightarrow$ AC $\leftrightarrow$ WO.

Proof. Our method is to show ZL $\rightarrow$ HMP $\rightarrow$ AC $\rightarrow$ WO $\rightarrow$ ZL.

(i) We show ZL $\rightarrow$ HMP. Let P be a partially ordered set, A a chain in P, and let $\mathscr{C}$ be the set of chains C in P with $A \subset C$. $\mathscr{C}$ is partially ordered by inclusion, and the union of a chain of chains is a chain. So $\mathscr{C}$ meets the hypothesis of Zorn's lemma. Hence $\mathscr{C}$ has a maximal element, and we are done.

(ii) We show HMP→AC. Given $\{X_i : i \in I\}$ where $I \neq \emptyset$ and each $X_i \neq \emptyset$, let $\mathscr{F} = \{f : f$ is a function, $\operatorname{dom} f \subset I$, and for each $i \in \operatorname{dom} f$, $f(i) \in X_i\}$. (You should check that $\mathscr{F}$ is in fact a set, i.e., is not too big.) Order $\mathscr{F}$ by inclusion. Let $\mathscr{C}$ be a maximal chain in $\mathscr{F}$. Then $\bigcup \mathscr{C} \in \mathscr{F}$. Let $f^* = \bigcup \mathscr{C}$ and suppose $i \notin \operatorname{dom} f^*$ for some $i \in I$. Let $a \in X_i$. Then $f^* \cup \{(i, a)\} = g \in \mathscr{F}$ and $g > f$ for all $f \in \mathscr{C}$, so $\mathscr{C}$ is not maximal, a contradiction. Hence $f^* \in \prod_{i \in I} X_i$.

(iii) We show AC→WO. Let x be a set and let $I = \mathscr{P}(x) - \{x\}$. For $y \in I$ let $z_y = x - y$. Let $g \in \prod_{y \in I} z_y$, and define $f : \mathscr{P}(x) \to x \cup \{x\}$ as follows: $f(y) = g(y)$ for $y \in I$; $f(x) = x$. We define a function h by induction: $h(0) = f(0)$. If we know $h \mid \alpha$, then $h(\alpha) = f(h[\alpha])$ if $h[\alpha] \neq x$; otherwise $h(\alpha) = \{x\}$.

Claim 1. For some α, $h(\alpha) = \{x\}$.

Proof. Otherwise let $Z = \{a \in x : \exists \alpha\, h(\alpha) = a\}$. Z is a set by separation. If $\beta < \alpha$, then $h(\beta) \in h[\alpha]$, so $h(\beta) \neq h(\alpha) \notin h[\alpha]$; hence h is 1-1 onto Z. Hence h^{-1} is a function from Z onto ON, which contradicts the axiom of substitution.

Claim 2. If α is the first ordinal where $h(\alpha) = \{x\}$, then $h \mid \alpha$ induces a well-ordering of x.

Proof. If $y \notin h[\alpha]$ and $y \in x$, then $h[\alpha] \neq x$, so $h(\alpha) \neq \{x\}$, a contradiction.

(iv) We show WO→ZL. Let P satisfy the hypothesis of ZL, and by WO let $P = \{p_\alpha : \alpha < \beta\}$ for some ordinal β. We define $f : \beta \to P$ by induction as follows: $f(\alpha) = p_\delta$ if δ is the least ordinal with $p_\delta > p$ for all $p \in f[\alpha]$. Otherwise $f(\alpha) = p_0$. Note that if $f(\alpha) = p_\delta$ then either $\delta = 0$ or $\delta \geq \alpha$. As in theorem 36, method I, $p_0 = f(0)$. Hence if $A = \{f(\alpha) : \alpha < \beta\}$, then A is a chain. By hypothesis, A has an upper bound p. Suppose p is not maximal in P. Then there is $q = p_\gamma > p$. But then $p_\gamma > r$ for all $r \in f[\gamma]$, so $q = f(\gamma)$, hence $q \in A$ and $q \leq p$, a contradiction.

SECTION 3.7. MODELS OF REGULARITY AND CHOICE

Definition 39. X is a model of regularity iff $\forall x \in X$ with $x \cap X \neq \emptyset$ there is some $y \in x \cap X$ with $y \cap x \cap X = \emptyset$.

Definition 40. Let X be transitive. X is a model of choice iff, for every

$I \in X$ and every set $\{x_i : i \in I\} \in X$ where each $x_i \neq \emptyset$, there is a function $f \in X$ where dom $f = I$ and, for each $i \in I$, $f(i) \in x_i$.

Note that, by transitivity, each x_i and $f(i)$ are elements of X, and $f \subset X$.

Theorem 41. Every set is a model of regularity.

   Proof. Given X and $x \in X$ with $x \cap X \neq \emptyset$, by regularity there is some y minimal in $x \cap X$; i.e., $y \cap (x \cap X) = \emptyset$. But this is just the y we are looking for.

Theorem 42. Let X be a transitive set closed under power set and union (i.e., if $x \in X$, then $\mathcal{P}(x) \in X$ and $\bigcup x \in X$) where, in addition, $a, b \in X$ implies $a \times b \in X$. Then X models choice.

   Proof. Given $I, \{x_i : i \in I\}$ both elements of X, note by the second and third hypotheses on X that $I \times \bigcup\{x_i : i \in I\} \in X$. If f is a choice function in $\Pi\{x_i : i \in I\}$, then $f \subset I \times \bigcup\{x_i : i \in I\}$. Hence, since X is closed under power set, by transitivity $f \in X$, and we are done.

Note that the proof of theorem 42 needed only weak closure under power set; if $x \in X$, then $\mathcal{P}(x) \subset X$.
   A word of caution: definitions 39 and 40 only provide models of one version of each axiom, e.g., without the other axioms of ZFC a model of AC need not be a model of WO.

EXERCISES FOR CHAPTER 3

1. Show that
   (a) If x is transitive, so are $S(x)$ and $x \cup \mathcal{P}(x)$.
   (b) If x_i is transitive for all $i \in I$, so is $\bigcup_{i \in I} x_i$.
   (c) If x, y are transitive, so is $x \cap y$.
   (d) There are transitive x, y with $x - y$ not transitive.
2. (a) What is the transitive closure of $X = \{0, 3, \{5, 7\}\}$?
   (b) What is the transitive closure of $X = \{x_n : n \in \mathbb{N}\}$ where $x_0 = \emptyset$ and each $x_{n+1} = \{x_n\}$?
3. (a) Show that each V_α is transitive.
   (b) Show that if $\alpha < \beta$ then $V_\alpha \subset V_\beta$.
   Do *not* use lemma 26.

4. Prove that if x is nonempty and transitive then $\emptyset \in x$.

5. Prove that for all ordinals α, rank $\alpha = \alpha$.

6. Show that if rank x = rank $y = \alpha$ then

   (a) rank$\{x, y\} = \alpha + 1$.

   (b) rank$(x, y) = \alpha + 2$.

   (c) rank$(x \cup y) = \alpha$.

7. Find x so rank $\bigcup x$ = rank x. Find x so rank $\bigcup x <$ rank x.

8. What is V_4?

9. (a) Show that every vector space has a basis, where a basis is a maximal linearly independent set.

   (b) Without using AC show that the vector space over $\mathbb{R}$ of polynomials with coefficients in $\mathbb{R}$ has a basis. (*Hint*: Define the basis.)

10. A *selector* for a family $\{X_i : i \in I\}$ is a 1–1 choice function.

   (a) Show that if, for each $n < \omega$, X_n has at least $n + 1$ elements then $\{X_n : n < \omega\}$ has a selector.

   (b) Let $\delta \in ON$. Show that if, for each $\alpha < \delta$, $\bigcup_{\beta < \alpha} X_\beta \neq X_\alpha$ then $\{X_\alpha : \alpha < \delta\}$ has a selector.

11. Formalize and prove the following statement: Given a drawer with infinitely many pairs of socks, you can pick out one sock from each pair.

12. We define a family of sets A to be pairwise disjoint if $a \cap b = \emptyset$ for all distinct $a, b \in A$. Show that if $B \subset \mathcal{P}(x)$ for some x then there is a maximal pairwise disjoint $A \subset B$; i.e., A is pairwise disjoint, $A \subset B$, and if $a \in B - A$ then $A \cup \{a\}$ is not pairwise disjoint.

13. We define a family of sets A to be linked if $a \cap b \neq \emptyset$ for all $a, b \in A$. Show that if $B \subset \mathcal{P}(x)$ for some x then there is a maximal linked $A \subset B$; i.e., A is linked, $A \subset B$, and if $a \in B - A$ then $A \cup \{a\}$ is not linked.

14. We define a family of sets A to be centered if every intersection of finitely many elements of A is nonempty. Show that if $B \subset \mathcal{P}(x)$ for some x then there is a maximal centered $A \subset B$; i.e., A is centered, $A \subset B$, and if $a \in B - A$ then $A \cup \{a\}$ is not centered.

4

THE FOUNDATION OF MATHEMATICS

INTRODUCTION

We have made the claim that all of mathematics can be done in the language of set theory,[†] i.e., that every mathematical statement can be translated into a formula whose only nonlogical symbol is $\in$. We have already substantiated some of this claim, in two ways: abstract structures such as linear orders have been defined in set-theoretic language, and concrete objects such as 0 and $\mathbb{N}$ have found representations as sets. The former is easier than the latter, and in this chapter it is the representation of concrete objects with which we are concerned. We will define set-theoretic representations of $\mathbb{Z}, \mathbb{Q}, \mathbb{R}$, of the arithmetic operations on each, of the order relation on each, and show that $\mathbb{R}$ has the desired geometric property of "no holes."

Twenty years ago, when the impact of modern set theory on the rest of mathematics was just beginning, this material was standard in foun-

---

[†]This claim depends, of course, on what you mean by mathematics. Intuitionists and finitists, who object to certain standard mathematical techniques, would declare set theory as we have presented it to be false, the former because it uses logical laws to which they object, such as the law of the excluded middle—either A holds or $\neg A$ holds; the latter because they declare infinite objects absurd. But the same charges can be made against much of standard mathematics. Less philosophical and more relevant are the objections of category theorists, who deal with objects that are classes—all groups, all rings—and then bind these together into classes of classes, etc. There is a set-theoretic trick to deal with this: Limit the cardinality of the objects you are working with; e.g., discuss only all groups of size $< \kappa$. But the category theorists respond, with some justification, that this is more than a tad artificial. There is no denying, however, that set-theoretic techniques have had powerful consequences in many fields of mathematics. This brief chapter seeks to develop the intuition behind this.

dations of mathematics courses, where it was taught in some detail in lieu of more set theory. Now, when there is so much set theory to learn, this older material is often left out entirely, making the root of set theory's profound impact on mathematics obscure. This chapter attempts a middle ground. To do all of this material in full detail would be tedious and would provide only the insight that it can be done. But this insight is itself a revolutionary one, and the reader deserves a demonstration of its reasonableness. That is the purpose of this chapter. Nearly all proofs are left to the exercises.

SECTION 4.1. ARITHMETIC ON $\mathbb{N}$

We already know what the elements of $\mathbb{N}$ are and how order is defined. The arithmetic operations are defined by induction:

$n +_N 0 = n$; $n +_N (m + 1) = (n +_N m) + 1$ (where $k + 1$ is just $S(k)$).

$n \cdot_N 0 = 0$; $n \cdot_N (m + 1) = (n \cdot_N m) +_N n$.

$n E_N 0 = 1$; $n E_N (m + 1) = n \cdot_N (n E_N m)$. Here $n E_N m$ is our notation for n^m.

In these definitions, the subscript $\mathbb{N}$ means that the operations are only defined on $\mathbb{N}^2$.

We can also add the definitions: $n -_N m = k$ iff $n +_N k = m$ and $n \div_N m = k$ iff $m \cdot_N k = n$.

Note that here, as throughout this chapter, our formulas are abbreviations of formulas whose only nonlogical symbol is $\in$. This point will not be made again.

What needs to be checked here is that these definitions do what they are supposed to; i.e., if $n +_N m = k$, then k is what we usually mean by $n + m$. Since our definitions are exactly the definitions we all learned in elementary school, this presents no problem.

SECTION 4.2. ARITHMETIC ON $\mathbb{Z}$

Our first instinct is to represent the elements of $\mathbb{Z}$ by ordered pairs (n, m) where $n, m \in \omega$ and (n, m) represents the integer $n - m$. But then the representation of integers is not unique: is -2 to be represented by $(0, 2), (1, 3), (57, 59) \dots$? So instead we use equivalence classes of ordered pairs of natural numbers: $(n, m) \equiv (n', m')$ iff $n +_N m' = n' +_N m$. Then $\mathbb{Z}$ is identified with $\{[(n, m)]: n, m \in \omega\}$.

We define arithmetic on $\mathbb{Z}$ as follows:

$$[(n, m)] +_z [(n', m')] = [(n +_z n', m +_N m')].$$
$$[(n, m)] -_z [(n', m')] = [(n, m)] +_z [(m', n')].$$
$$[(n, m)] \cdot_z [(n', m')] = [(((n \cdot_N n') +_N (m \cdot_N m')),$$
$$((n \cdot_N m') +_N (m \cdot_N n')))].$$

Note that, in $\mathbb{Z}$, 0 is represented by $\{(n, n): n \in \omega\} = 0_z$.

Again, if $a, b \in \mathbb{Z}$ and $b \neq 0_z$, we can define $a \div_z b = c$ iff $b \cdot_z c = a$.

We define order on $\mathbb{Z}$: $[(n, m)] \leq_z [(n', m')]$ iff $n +_N m' \leq_N n' +_N m$.

Now we have two sorts of questions to answer: Do our definitions do what they are supposed to do, and, since $\mathbb{Z}$ is a set of equivalence classes, are they well defined, e.g., if $[(n, m)] = [(n', m')]$ and $[(k, j)] = [(k', j')]$, does $[(n, m)] +_z [(k, j)] = [(n', m')] +_z [(k', j')]$? The reader will be asked to check some of this in the exercises.

SECTION 4.3. ARITHMETIC ON Q

Again, $\mathbb{Q}$ is defined to be a set of equivalence classes of ordered pairs. This time, the ordered pairs are in $\mathbb{Z}^2$, and (a, b) is supposed to be a representative of a/b. So let $a, b, a', b' \in \mathbb{Z}$. We say $(a, b) \equiv (a', b')$ iff $a \cdot_z b' = a' \cdot_z b$. We define $\mathbb{Q} = \{[(a, b)]: a, b \in \mathbb{Z} \text{ and } b \neq 0_z\}$.

For readability, if $a, b \in \mathbb{Z}$ we write ab instead of $a \cdot_z b$.

We will define the operation of addition on $\mathbb{Q}$ and the relation of order on $\mathbb{Q}$, leaving the rest to the reader:

$$[(a, b)] +_Q [(c, d)] = [((ad +_z cb), bd)].$$

$[(a, b)] \leq_Q [(c, d)]$ iff $b \geq_z 0$, $d \geq_z 0$, and $ad \leq_z cb$, where 0 is understood to be 0_z.

Again, we have to ask if our definitions are well-defined, and if they do what they are supposed to do.

Note that 0 is represented in $\mathbb{Q}$ as $0_Q = \{(a, b) \in \mathbb{Q} \mid a = 0_z\}$, i.e., as a set of pairs of equivalence classes of pairs.

SECTION 4.4. ARITHMETIC ON R

The previous three sections have been primarily concerned with the algebraic properties of $\mathbb{N}$, $\mathbb{Z}$, and $\mathbb{Q}$, with order a secondary concern, definable from the algebra. Given the fact that $\mathbb{Z}$ is an algebraic extension of $\mathbb{N}$, and $\mathbb{Q}$ an algebraic extension of $\mathbb{Z}$, this is what we would expect. But $\mathbb{R}$ is different. It adds elements that no simple equation with coefficients in $\mathbb{Q}$ can account for. And it has a crucial geometric property as an order that cannot be derived from its algebraic structure: $\mathbb{R}$ is a continuum; that is, it has no holes. This property of having no holes is formalized as the least upper bound property: Every bounded subset of $\mathbb{R}$ has a least upper bound. How can we represent $\mathbb{R}$ in terms of $\mathbb{Q}$ and still capture our intuition of $\mathbb{R}$'s geometric completeness? This was an important question in nineteenth century mathematics, one of the keys to making the notions of calculus precise. More than one formal solution was offered; the one which is easiest to work with (all of the solutions are provably equivalent) is the method of Dedekind cuts.

Definition 1. A subset A of $\mathbb{Q}$ is said to be a Dedekind cut iff

(a) A is a proper initial segment of $\mathbb{Q}$, and
(b) A has no largest element.

$\mathbb{R}$ is defined to be the set of Dedekind cuts. Note that the union of Dedekind cuts is either a Dedekind cut or $\mathbb{Q}$.

The instinct here is that $\sqrt{2}$ is represented by $\{q \in \mathbb{Q}: q < \sqrt{2}\}$. Property (b) is to ensure that each real number has only one representative: We do not want both $\{q \in \mathbb{Q}: q < 17\}$ and $\{q \in \mathbb{Q}; q \leq 17\}$ to be elements of $\mathbb{R}$.

Defining addition is easy: $A +_{\mathbb{R}} B = \{a +_{\mathbb{Q}} b: a \in A \text{ and } b \in B\}$.
For subtraction: $A -_{\mathbb{R}} B = \{q: \forall b \in B \, \exists a \in A \, (q <_{\mathbb{Q}} a -_{\mathbb{Q}} b)\}$.
Defining multiplication on the reals is left to the exercises.
Order is easy: $A \leq_{\mathbb{R}} B$ iff $A \subset B$. This is a linear order: Initial segments of a linear order (and $\mathbb{Q}$ is linearly ordered) are either equal or one is a proper subset of the other.

What about the least upper bound property? Suppose $\mathscr{A}$ is a collection of Dedekind cuts. If $\mathscr{A}$ is bounded, then there is some Dedekind cut B where $A \subset B$ for all $A \in \mathscr{A}$. But then $\bigcup \mathscr{A}$ cannot be $\mathbb{Q}$, so $\bigcup \mathscr{A}$ is a Dedekind cut. $\bigcup \mathscr{A}$ is clearly an upper bound for $\mathscr{A}$. We show it is the least upper bound: If $C \supset A$ for all $A \in \mathscr{A}$, then $C \supset \bigcup \mathscr{A}$, hence $C \geq_{\mathbb{R}} \bigcup \mathscr{A}$.
Now we must ask if our definitions make sense (e.g., is $A +_{\mathbb{R}} B$ a

Dedekind cut?) and if they do what they are supposed to do. Again, this is left to the exercises.

Note that the representation of 0 in $\mathbb{R}$ is $\{q \in \mathbb{Q}: q <_\mathbb{Q} 0_\mathbb{Q}\}$, i.e., a set of equivalence classes of pairs of equivalence classes of pairs of natural numbers.

EXERCISES FOR CHAPTER 4

1. Show that our definition of multiplication on $\mathbb{Z}$ does what it is supposed to and that the order we defined on $\mathbb{Z}$ is linear.

2. Show that our definition of addition on $\mathbb{Z}$ is well-defined.

3. Define multiplication on $\mathbb{Q}$, and show that it is well-defined.

4. Show that if A and B are Dedekind cuts, so is $A +_\mathbb{R} B$.

5. Show that our definition of subtraction works.

6. Define multiplication on $\mathbb{R}$ by cases: Both numbers are nonnegative; exactly one number is negative; both numbers are negative.

5

INFINITE NUMBERS

INTRODUCTION

Before Zermelo, before Fraenkel, before Frege defined the number zero in nearly ninety closely argued pages, there was Cantor.[†]

Cantor's achievement was monumental: He made the study of infinity precise. He discovered infinite ordinals, he discovered the variety of infinite cardinals, and he invented three of the most useful tools in mathematical logic: dove-tailing (the ancestor of time-sharing), diagonalization, and the back-and-forth argument (used in model theory). His work went so counter to the ideas of the time—infinity is out there were you cannot get at it, there is only one infinity, and nothing infinite can truly be said to exist anyway—that he was bitterly fought by philosophers, mathematicians, and even theologians (who associated infinity with God and hence saw Cantor as sacrilegious). But even his greatest critics had to admit that the only way to seriously deny the validity of his work, so rooted is it in mathematical practice, is to deny even that $\mathbb{N}$ is a set; i.e., to insist that only finite objects exist, as Aristotle had. Cantor's great enemy Kronecker tried to convince other mathematicians that the infinite could not be spoken of, but the effort failed. In 1925 David Hilbert acknowledged the general acceptance of Cantorian infinity by saying "no one can drive us from the paradise that Cantor created for us."

We have already talked about infinite ordinals a bit. In this chapter we give some basic facts about cardinality and then do some ordinal arithmetic, some cardinal arithmetic, and some more advanced work with cardinals. For the first two sections of this chapter we will point out all

[†]Cantor's pioneering work was done in the 1870s and 1880s; Frege's *Foundations of Arithmetic* was written at the turn of the century; Zermelo's seminal paper was in 1908.

uses of the axioms of choice. After that we simply assume that we are working in full ZFC—otherwise there is little that can be proved.

SECTION 5.1. CARDINALITY

Rather than defining a cardinal number (a task fraught with difficulties) we will define cardinality. We write $|A| \leq |B|$ for "the cardinality of A is smaller than or equal to the cardinality of B" and $|A| = |B|$ for "A and B have the same cardinality." We define these forthwith:

Definition 1. $|A| \leq |B|$ iff there is a 1–1 function $f: A \to B$. $|A| = |B|$ iff there is a 1–1 function from A onto B; $|A| < |B|$ iff $|A| \leq |B|$ and $|A| \neq |B|$.

In particular, if $A \subset B$, then $|A| \leq |B|$.

Definition 1 defines a relation between A and B. It does not define an object called "the cardinality of A." This phrase is just how our language works, a quirk of grammar.

Why is definition 1 reasonable? You have a room with chairs. Suppose everyone in the room is sitting down. If there are some chairs left over, we know without counting that there are more chairs than people. If there are no empty chairs, we know without counting that there are the same number of chairs and people. Suppose every chair is full and there are some people left standing. Then we know without counting that there are more people than chairs. The chairs can be in rows, in a single line, in a circle, arranged as in a concert hall—it does not matter. We do not care about the arrangement, only whether every person has a chair or every chair a person.

Suppose, for every set A, we had defined a set $|A|$ satisfying definition 1. (We will actually do this under AC.) Then the relation $\leq$ would be reflexive and transitive. In this loose way of speaking, the next theorem shows antisymmetry.

Theorem 2 (Schröder–Bernstein). If $|A| \leq |B|$ and $|B| \leq |A|$, then $|A| = |B|$.

At the end of this section we will show how easy it is to prove the Schröder–Bernstein theorem using AC. Without AC it takes some care.

Proof of the Schröder–Bernstein Theorem. We are given $f: A \to B$ and $g: B \to A$; both f and g are 1–1. Our task is to find some 1–1 h from A onto B.

By induction on ω let us construct descending sequences $\{A_n: n < \omega\}$ and $\{B_n: n < \omega\}$ where $A = A_0$, $B = B_0$, each $A_{n+1} \subset A_n$, each $B_{n+1} \subset B_n$, and if $n > 0$ we let $A_n = g[B_{n-1}]$ and $B_n = f[A_{n-1}]$:

Let $A_n^* = A_n - A_{n+1}$, $B_n^* = B_n - B_{n+1}$. Let $A^\dagger = \bigcap_{n<\omega} A_n$, $B^\dagger = \bigcap_{n<\omega} B_n$.

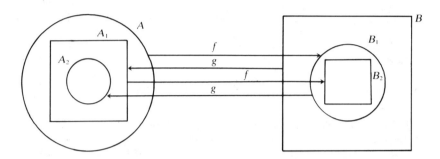

Claim 1. For all $n < \omega$, $f \mid A_n^*$ is onto B_{n+1}^* and $g \mid B_n^*$ is onto A_{n+1}^*.

Proof. Since $B_{n+1} = f[A_n]$ and $B_{n+2} = f[A_{n+1}]$ and since f is 1-1, we conclude that $f[A_n - A_{n+1}] = B_{n+1} - B_{n+2}$. A similar proof works for g.

Claim 2. $A = A^\dagger \cup \bigcup_{n<\omega} A_n^*$; $B = B^\dagger \cup \bigcup_{n<\omega} B_n^*$.

Proof. If $x \in B - B^\dagger$, then there is some n with $x \notin B_n$. Let k be the least such. Then $k > 0$ since $B = B_0$, so $x \in B_{k-1}^*$. A similar proof works for A.

Claim 3. $\{A^\dagger\} \cup \{A_n^*: n < \omega\}$ is pairwise disjoint, as is $\{B^\dagger\} \cup \{B_n^*: n < \omega\}$.

Proof. Since if $m < n$ then $A_n \subset A_{m+1}$, distinct A_n^*'s are disjoint. And $x \notin \bigcup_{n<\omega} A_n^*$ if $x \in A^\dagger$ by definition of $A^\dagger$. A similar proof works for B.

Claim 4. $f \mid A^\dagger$ is onto $B^\dagger$.

Proof. Since $B^\dagger \subset B_1 = f[A]$, $B^\dagger$ is a subset of range f. Since each $f \mid A_n^*$ is onto B_{n+1}^* and f is onto B_1, $B^\dagger \subset f[A - \bigcup_{n<\omega} A_n^*]$. So by claims 2 and 3, $B^\dagger = f[A^\dagger]$.

We are ready to construct a 1-1 function h from A onto B as follows: $h \mid A^\dagger = f$; for all $n < \omega$, $h \mid A_{2n} = f \mid A_{2n}$ and $h \mid A_{2n+1} = g^{-1} \mid A_{2n+1}$. By claims 1 and 4, h is onto. Since f and g are 1-1, by claim 3, h is 1-1. So we are done.

This theorem justifies using the word "size" in connection with cardinality.

Two useful lemmas are

Lemma 3. Assume AC. If some f takes a set A onto a set B, $|A| \geq |B|$.

Proof. For $b \in B$, let $x_b = f^{-1}(b)$. Let $h \in \prod_{b \in B} x_b$. Since the x_b's are disjoint, h is 1–1. So $|B| \leq |\text{range } h|$. Since range $h \subset A$, we are done.

Lemma 4. Suppose $|X| \leq |Y|$. Let $x \in X$, $y \in Y$. Then $|X - \{x\}| \leq |Y - \{y\}|$.

Proof. Let $f: X \to Y$ be 1–1. Define $f^*: X - \{x\} \to Y - \{y\}$ as follows: $f^*(z) = f(z)$ if $z \neq x$ and $f(z) \neq y$. If $z \neq x$ and $f(z) = y$, let $f^*(z) = f(x)$. The reader can easily check that f^* is 1–1 and $\text{dom}(f^*) = X - \{x\}$.

Recall our earlier definition of finite ordinal. Using the terminology of this section, a set X is finite iff, for some $n \in \omega$, $|X| = |n|$; a set which is not finite is called infinite. There is another useful characterization of finite, equivalent under AC.

Theorem 5

  (a) If X is finite, then $|X| \neq |Y|$ for any proper subset Y of X.
  (b) Assume AC. If $|X| \neq |Y|$ for any proper subset Y of X, then X is finite.

Proof. For (a), by an argument similar to lemma 4, it suffices to show that if $n \in \omega$ and $m < n$ then $|m| < |n|$. This is immediate, since $m < n$ iff $m \subset n$.

For (b), notice that WO implies that every infinite set has a copy of ω embedded in it. So let X be infinite, and let $f: \omega \to X$ be 1–1. Then let $g: X - \{f(0)\} \to X$ be defined by $g(a) = a$ if $a \notin \text{range } f$; $g(f(n+1)) = f(n)$. Since g is 1–1 and onto, X has the same cardinality as a proper subset of itself, and we are done.

A set X satisfying "$|X| \neq |Y|$ for any proper subset Y of X" is called Dedekind-finite. Without AC there can be sets which are Dedekind-finite but not finite.

Cardinalities can be classified in many ways. Our first, crude, attempt was to classify them as finite or infinite. Our second, slightly more subtle, attempt is

Definition 6. A set X is countable iff $|X| \leq |\omega|$; uncountable iff not countable. X is denumerable iff $|X| = \omega$.

Cantor's two great early theorems on cardinality were that the rationals are countable but the reals are not.

Theorem 7. $\mathbb{Q}$ is countable.

We give two proofs.

Proof 1. To each nonzero $q \in \mathbb{Q}$ we associate the unique pair (n_q, m_q), where $q = n_q / m_q$, $m_q > 0$, and n_q, m_q have no common divisors except 1; if $q = 0$, let $n_q = 0$, $m_q = 1$. Let r, s, t be distinct prime numbers and define $f(q) = r^{n_q} \cdot s^{m_q}$ if q is positive; $f(q) = t^{|n_q|} \cdot s^{m_q}$ otherwise. By arithmetic, f is 1-1. Since $\mathbb{N}$ embeds in $\mathbb{Q}$, by the Schröder–Bernstein theorem we are done.

Cantor did not come up with this proof—the Schröder–Bernstein theorem had not been proved yet. Instead he came up with the following proof, in which he invented a technique known as dove-tailing, which is the basic principle behind time-sharing. Because this method of proof is so important, we give it here.

Proof 2. Let n_q, m_q be as in the first proof. We construct the function $f: \omega \to \mathbb{Q}$ as follows: $f(0) = 0$. (Note that $n_0 = 0$, $m_0 = 1$). Suppose we know $f(k) = q_k$. Let $n = n_{q_k}$, $m = m_{q_k}$, and let $r = |n| + m$.

Case 1. k is odd. Then $f(k+1) = -q_k$.

Case 2. k is even. If $m > 1$ and there is j, $m > j \geq 1$ with j, $r - j$ having no common factor except 1, pick the greatest such j where $(r-j)/j \notin$ range $f \upharpoonright k+1$ and let $f(k+1) = (r-j)/j$. Otherwise let $f(k+1) = 1/r$.

Here is a picture that explains the proof for even k:

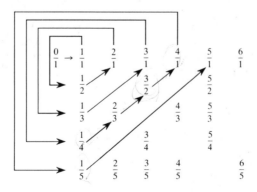

Notice that the gaps are exactly fractions not in reduced form (e.g., $\frac{2}{2}, \frac{4}{2}, \frac{3}{3}, \frac{2}{4} \ldots$). Notice that what we really have is an infinite number of tasks ("list every element in this row") which we finish by doing a little on one, a little on another, then back to the first, now on to the third, back to the second, to the first, and so on, i.e., infinite time-sharing, also known as dove-tailing.

With AC, a corollary to the proof of theorem 7 is that the countable union of countable sets is countable.

The reason this proof does not generalize without AC is that we relied heavily on the fact that we have simultaneously at hand a well-ordering of each row. The generalization of theorem 7—if Y is infinite, and each $|y| = |Y|$ for $y \in Y$, then $|\bigcup Y| = |Y|$—is not a theorem of ZF, even if Y is countable. There are models of ZF in which it fails, although it holds in ZFC.

The next theorem is essentially Cantor's second proof that the reals are uncountable. (The first proof was analytic and did not generalize.) The method is called diagonalization and is a powerful technique in set theory and logic, used, for example, to prove Gödel's incompleteness theorems, and frequently invoked in theoretical computer science.

Theorem 8. For all x, $|x| < |\mathcal{P}(x)|$.

Proof. It is trivial to show that $|x| \le |\mathcal{P}(x)|$—just let $f(y) = \{y\}$. What we need to show is that $|x| \neq |\mathcal{P}(x)|$.

Suppose f is a function from x to $\mathcal{P}(x)$. We construct a set $y \subset x$ with $y \notin$ range f as follows: $a \in y$ iff $a \notin f(a)$. If, for some a, $y = f(a)$, then $a \in y$ iff $a \notin y$, a contradiction.

The brevity of the above proof should not obscure its ingeniousness. The reader interested in how Cantor came upon it, and the mixed reception to the theorem, should read one of the historical books listed in the bibliography.

It is useful to note that $|\mathcal{P}(x)| = |^x 2|$ as follows: Define $\chi: \mathcal{P}(x) \to {}^x 2$ by $\chi(y)(a) = 0$ if $a \in y$; $\chi(y)(a) = 1$ if $a \notin y$. (We write $\chi(y) = \chi_y =$ the characteristic function of y.) It is easy to show that χ is 1–1 and onto. Hence, by theorem 8, each $|x| < |^x 2|$.

Corollary 9

(a) There is no biggest set: For all x there is some y, $|y| > |x|$.

(b) $\mathbb{R}$ is uncountable.

Proof. (a) is clear. Let us do (b).

Let A be the set of reals between 0 and 1 which, in their decimal expansions, have only 0's and 1's, e.g., .101001000.... Then $A \subset \mathbb{R}$, so $|A| \le |\mathbb{R}|$. But $|A| = |^\omega 2|$, since each element of A corresponds to a unique sequence of 0's and 1's. So $|\omega| < |A| \le |\mathbb{R}|$.

It is time to face the question "what is a cardinal number?" The reply of the early set theorists was "$|X|$ is the class of all Y with $|Y| = |X|$." This is highly unsatisfactory: We want to define numbers as canonical objects and not as equivalence classes; if forced to deal with equivalence classes, we would at least want to cut them down to sets and not proper classes. This second objection can be dealt with—define $|X|$ to be the set of all Y of minimal rank so $|Y| = |X|$—but the first objection remains. Without AC there is no useful way to define cardinal numbers.

On the other hand, there is a class of ordinal numbers that look like good candidates for canonical cardinals.

Definition 10. An ordinal κ is an initial ordinal iff, for all $\alpha < \kappa$, $|\alpha| < |\kappa|$.

For example, ω is an initial ordinal; $\omega + 1$ is not (since the map $f(0) = \omega$, $f(n+1) = n$ is 1-1 from ω onto $\omega + 1$). Similarly, if $\alpha \ge \omega$, then $|\alpha| = |\alpha + 1|$.

By a cardinal number we will mean an initial ordinal. Thus every infinite cardinal is a limit ordinal.

Theorem 11. AC is equivalent to "every x is isomorphic to a unique cardinal number."

Proof. If AC fails, then some x is isomorphic to no ordinal, hence certainly not to an initial one. If AC holds, then by WO for each x there is an ordinal α with $|x| = |\alpha|$. Let $A = \{\gamma \le \alpha : |\gamma| = |x|\}$. Let $\kappa = \inf A$. If $|\kappa| = |\beta|$ for some $\beta < \kappa$, then $\beta \in A$, which is a contradiction. So κ is an initial ordinal, and we are done.

Under AC, then, we can define the cardinality of a set.

Definition 12. Assume AC. Then $|x| = \kappa$ iff κ is the unique cardinal number for which $|x| = |\kappa|$.

The proof of the Schröder–Bernstein theorem under AC is now trivial: If $|A| \le |B| \le |A|$ and $\kappa = |A|$, $\lambda = |B|$, then $\kappa \le \lambda \le \kappa$, so $\kappa = \lambda$, and we are done.

Let us introduce some cardinals: $\omega_0 = \omega$; ω_1 is the first uncountable cardinal; in general if $\alpha \in$ ON, then $\omega_{\alpha+1}$ is the first cardinal greater than ω_α; if α is a limit, then $\omega_\alpha = \bigcup_{\beta<\alpha} \omega_\beta$. (This is indeed a cardinal. Otherwise $|\omega_\alpha| = \omega_\beta < \omega_{\beta+1}$ for some $\beta < \alpha$, but $\omega_{\beta+1} = |\omega_{\beta+1}| \le |\omega_\alpha|$, which contradicts $\omega_\beta < \omega_{\beta+1}$.) ω_α is sometimes written as $\aleph_\alpha$ when its cardinal nature is being emphasized. ($\aleph$ is the Hebrew letter aleph.) Notice that, since as an ordinal $\omega_\alpha = \{\beta: \beta < \omega_\alpha\}$, each $\omega_\alpha = \{\beta \in$ ON$: |\beta| < \omega_\alpha\}$.

Assuming AC we write $2^\kappa = |{}^\kappa 2|$ and define the $\beth_\alpha$'s ($\beth$ is the Hebrew letter bet without its diacritical mark): $\beth_0 = \omega$; $\beth_{\alpha+1} = 2^{\beth_\alpha}$ for all ordinals α; if α is a limit ordinal, then $\beth_\alpha = \bigcup_{\beta<\alpha} \beth_\beta$. We will not use the bet notation, but many authors do.

Note that the definition of the $\aleph_\alpha$'s does not depend on AC.

If $\lambda = \aleph_{\alpha+1}$ and $\kappa = \aleph_\alpha$, we say $\lambda = \kappa^+$ and λ is a successor cardinal. If $\lambda = \aleph_\alpha$ and α is a limit ordinal, we say λ is a limit cardinal. A cardinal κ is infinite iff $\omega \le \kappa$.

SECTION 5.2. ORDINAL ARITHMETIC

Line up three tortoises. Then line up five more behind them. Eight tortoises are now lined up. This is how we define $\alpha + \beta$: a copy of α followed by a copy of β.

Line up three tortoises. Line up three more behind them. And again three more. And again. Three times four tortoises are now lined up. This is how we define $\alpha \cdot \beta$: (α followed by α followed by α ...) β times.

The formal definitions are inductive.

Definition 13. For all ordinals α, β: $\alpha + 0 = \alpha$; $\alpha + (\beta + 1) = (\alpha + \beta) + 1$; if β is a nonzero limit, then $\alpha + \beta = \bigcup_{\gamma<\beta} (\alpha + \gamma)$.

Definition 14. For all ordinals α, β: $\alpha \cdot 0 = 0$; $\alpha \cdot (\beta + 1) = (\alpha \cdot \beta) + \alpha$; if β is a limit then $\alpha \cdot \beta = \bigcup_{\gamma<\beta} (\alpha \cdot \gamma)$.

Note that the nonlimit clauses exactly parallel the definitions in $\mathbb{N}$. Here is a picture of $\alpha + \beta$:

Here is a picture of $\alpha \cdot \beta$:

For completeness we include

Definition 15. For all ordinals α, β: $\alpha^0 = 1$; $\alpha^{\beta+1} = \alpha^\beta \cdot \alpha$; if β is a nonzero limit, then $\alpha^\beta = \bigcup_{\gamma<\beta} \alpha^\gamma$.

Notice that addition and multiplication do not commute: $2 + \omega = \sup\{2 + n: n < \omega\} = \omega < \omega + 1 < \omega + 2$; $2 \cdot \omega = \sup\{2n: n < \omega\} = \omega < \omega + 1 < \omega + \omega = \omega \cdot 2$. We must also be careful not to confuse ordinal exponentiation with cardinal exponentiation: In ordinal exponentiation $2^\omega = \sup\{2^n: n < \omega\} = \omega$, but in cardinal exponentiation $2^\omega > \omega$. Unless otherwise noted, in this section all arithmetical operations are the ordinal operations. Subsequent sections use mostly cardinal operations, unless noted otherwise, with the exception that $\alpha + 1$ is always the ordinal successor of α.

While commutativity does not hold, associativity and left-distributivity do.

Theorem 16. For all ordinals α, β, γ,

(a) $(\alpha + \beta) + \gamma = \alpha + (\beta + \gamma)$
(b) $(\alpha \cdot \beta) \cdot \gamma = \alpha \cdot (\beta \cdot \gamma)$
(c) $\alpha \cdot (\beta + \gamma) = \alpha \cdot \beta + \alpha \cdot \gamma$

Proof. We prove (a); (b) and (c) are left to the reader. The proof of (a) is by induction on γ. Suppose it is true for all α, β and all $\delta < \gamma$.

CASE 1. $\gamma = \delta + 1$ for some δ. Then

$$(\alpha + \beta) + \gamma = (\alpha + \beta) + (\delta + 1)$$
$$= [(\alpha + \beta) + \delta] + 1 \quad \text{by definition of addition}$$
$$= [\alpha + (\beta + \delta)] + 1 \quad \text{by induction hypothesis}$$
$$= \alpha + [(\beta + \delta) + 1] \quad \text{by definition of addition}$$
$$= \alpha + [\beta + (\delta + 1)] \quad \text{by definition of addition}$$
$$= \alpha + (\beta + \gamma)$$

CASE 2. γ is a limit. We shall need the following claim, whose proof is left to the reader: For all ordinals α and sets of ordinals A, $\sup\{\alpha + \beta: \beta \in A\} = \alpha + \sup\{\beta: \beta \in A\}$. Given the claim:

$$(\alpha + \beta) + \gamma = \sup\{(\alpha + \beta) + \delta: \delta < \gamma\}$$
$$= \sup\{\alpha + (\beta + \delta): \delta < \gamma\} \quad \text{by induction hypothesis}$$

$$= \alpha + \sup\{\beta + \delta : \delta < \gamma\} \quad \text{by the claim}$$

$$= \alpha + (\beta + \gamma) \quad \text{by definition of addition}$$

Here is a picture of what is going on:

Given: α: _____

β: _ _ _ _ _ _ _ _ _ _ _ _ _

γ: . . . .

we have $(\alpha + \beta) + \gamma$: (_____ _ _ _ _ _ _ _ _ _ _ _ _ _) . . . .

and $\alpha + (\beta + \gamma)$: _____ (_ _ _ _ _ _ _ _ _ _ _ _ . . . .)

The picture for left distributivity is

$\alpha \cdot (\beta + \gamma)$: ―― ―― ―― . . . ―― ―― . . .

$\underbrace{\qquad\qquad\qquad}$

$\beta + \gamma$

$(\alpha \cdot \beta) + (\alpha \cdot \gamma)$: ―― ―― . . . ―― ―― . . .

$\underbrace{\qquad}\quad\underbrace{\qquad}$

$\beta \qquad\qquad \gamma$

Right distributivity fails for ordinal arithmetic. For example, $(\omega + 1) \cdot \omega = \omega \cdot \omega$. Here is a picture:

$\omega + 1$: ------- . . . -

$(\omega + 1) \cdot \omega$: ------ . . . - ------- . . . - ----- . . . - ------ . . . -

$\omega \cdot \omega$: ------ . . . ------- . . . ------- . . . ------ . . . -

But $(\omega \cdot \omega) + \omega > (\omega \cdot \omega) + 1 > \omega \cdot \omega$.

Let us give some concrete pictures of some small infinite ordinals.

Definition 17. A well-ordered set has order type α iff the set is order-isomorphic to the ordinal α (where we say the set X under the order $\leq$ is order-isomorphic to α iff there is a 1–1 onto function $f: X \to \alpha$ where $x \leq y$ iff $f(x) \in f(y)$).

For example, under the lexicographic ordering:

(i) $[\{0\} \times \alpha] \cup [\{1\} \times \beta]$ has order type $\alpha + \beta$.

(ii) $\beta \times \alpha$ has order type $\alpha \cdot \beta$.

(iii) $\gamma \times \beta \times \alpha$ has order type $\alpha \cdot \beta \cdot \gamma$.

The assertions (i), (ii), and (iii) are easily proved by induction.

Now let us find sets of reals with desired order types.

Example 18. We define numbers x_n, $x_{n,m}$, and $x_{n,m,k}$ for $n, m, k \in \omega$ as follows:

$$x_n \quad \text{is} \quad \underbrace{.1 \ldots 1}_{n \text{ ones}} 000 \ldots$$

$$x_{n,m} \quad \text{is} \quad \underbrace{.1 \ldots 1}_{n \text{ ones}} 0 \underbrace{1 \ldots 1}_{m \text{ ones}} 000 \ldots$$

$$x_{n,m,k} \quad \text{is} \quad \underbrace{.1 \ldots 1}_{n \text{ ones}} 0 \underbrace{1 \ldots 1}_{m \text{ ones}} 0 \underbrace{1 \ldots 1}_{k \text{ ones}} 000 \ldots$$

where the notation "$000 \ldots$" means a final string of zeros. Then

(a) $\{x_n : n \in \omega\}$ has order type ω.
(b) $\{x_{n,m} : n, m \in \omega\}$ has order type $\omega \cdot \omega$.
(c) $\{x_{n,m,k} : n, m, k \in \omega\}$ has order type $\omega \cdot \omega \cdot \omega$.

Proof. We prove (b) and leave (a) and (c) to the reader. By (ii), it suffices to show that $\{x_{n,m} : n, m \in \omega\}$ is order–isomorphic to $\omega \times \omega$ under the lexicographic order. That is, we must show that if $(n, m) \leq_L (k, j)$ then $x_{n,m} \leq x_{k,j}$. If $(n, m) = (k, j)$, we are done, so we are left with two cases.

CASE 1. $n < k$. Then the ith decimal place of $x_{n,m} =$ the ith decimal place of $x_{k,j}$ for $i \leq n$. But the $(n+1)$st decimal place of $x_{n,m}$ is 0, and the $(n+1)$st decimal place of $x_{k,j}$ is 1, so $x_{n,m} < x_{k,j}$.

CASE 2. $n = k$, $m < j$. Then the ith decimal place of $x_{n,m} =$ the ith decimal place of $x_{k,j}$ for $i \leq n + m + 1$. But, if $i = n + m + 2$, the ith decimal place of $x_{n,m} = 0$, and the ith decimal place of $x_{n,j} = 1$. So $x_{n,m} < x_{n,j}$.

The exercises expand on this method of embedding countable ordinals into $\mathbb{R}$.

The next theorem says that every well-ordered set has an order-type.

Theorem 19. Every well-ordered set is order-isomorphic to an ordinal.

Proof. Let X be well-ordered by $\leq$. We define f by induction on ON as follows: $f(0)$ is the least element in X. If $f(\alpha)$ is defined and in X, then $f(\alpha + 1)$ is the least element in X strictly greater than $f(\alpha)$, if such exists; otherwise $f(\alpha + 1) = X$. If α is a limit and $f(\beta)$ is defined and in X for all $\beta < \alpha$, then $f(\alpha)$ is the least element in X strictly greater than all $f(\beta)$, $\beta < \alpha$, if such exists; otherwise $f(\alpha) = X$.

Let γ be the first ordinal for which $f(\gamma) = X$. Then $f \mid \gamma$ is order-preserving by definition. If $f \mid \gamma$ is not onto X, let x be the least element in X with $x \notin f[\gamma]$. Let $\delta = \sup\{\alpha : f(\alpha) < x\}$. Then $f(\delta) = x$ and $\delta < \gamma$ by definition of γ. So $x \in f[\gamma]$, a contradiction.

We mix cardinality and ordinality with

Theorem 20. An ordinal α is countable iff some subset of $\mathbb{R}$ has type α.

Proof. If $X \subset \mathbb{R}$ has type α, then $X = \{x_\beta : \beta < \alpha\}$ where if $\beta < \gamma < \alpha$ then $x_\beta < x_\gamma$. Consider the open intervals $I_\beta = (x_\beta, x_{\beta+1})$ in $\mathbb{R}$. (If $\alpha = \beta + 1$, let $x_{\beta+1} = \infty$.) The I_β's are pairwise disjoint and each contains a rational, so there are only countably many of them. The function $f : X \to \mathcal{P}(\mathbb{R})$ given by $f(x_\beta) = I_\beta$ is 1–1 and onto $\{I_\beta : \beta < \alpha\}$, so X is countable.

Before proving the other direction we need

Lemma 21. A nonzero countable limit ordinal is some $\bigcup_{n < \omega} \beta_n$ where if $n < m$ then $\beta_n < \beta_m$.

Proof. Lemma 21 is equivalent to the statement: For every countable limit ordinal α there is an order-preserving map $f : \omega \to \alpha$ with $\{\gamma : \exists \beta > \gamma$ $(\beta \in \text{range } f)\} = \alpha$. Suppose α is a countable limit. Let $h : \omega \to \alpha$ be 1–1 and onto. We define $f : \omega \to \alpha$ as follows: Let $f(0) = h(0)$. Given $f(n)$ let k be the least ordinal greater than n with $h(k) > f(n)$ (why does k exist?). Let $f(n + 1) = h(k)$. By construction, f is order-preserving and $\{\gamma : \exists \beta > \gamma \ (\beta \in \text{range } f)\} = \alpha$. So we are done.

Now we return to the proof of theorem 20. We show by induction that if α is countable then some subset of $\mathbb{R}$ has type α. Suppose it is true for each $\beta < \alpha$. If $\alpha = \beta + 1$, then, since $\mathbb{R}$ is order-isomorphic to the open interval $(0, 1)$, some subset Y of $(0, 1)$ is of type β. Then $Y \cup \{1\}$ is of type α. If α is a limit, since α is countable, $\alpha = \bigcup_{n < \omega} \beta_n$ where we can assume that if $n < m$ then $\beta_n < \beta_m$. Let δ_n be the order type of $\{\gamma : \beta_n \leq \gamma < \beta_{n+1}\}$, let I_n be the open interval $(n, n + 1)$ in $\mathbb{R}$, and let $Y_n \subset I_n$ be of type δ_n. Then $\bigcup_{n < \omega} Y_n$ has type α.

Theorem 20 tells us why uncountable ordinals are so hard to picture: They cannot be embedded in $\mathbb{R}$. Note that the second part of the proof seems to use a mild form of choice, when we picked the Y_n's simultaneously. In fact, choice is not needed—by taking a bit more care the sets of reals we construct can be made definable.

Here are some examples of ordinal arithmetic equalities and inequalities involving uncountable ordinals:

Example 22

(a) $\omega + \omega_1 = \omega_1$.

(b) $\omega \cdot \omega_1 = \omega_1$.

(c) $\omega^{\omega_1} = \omega_1$.

(d) $\omega_1 < \omega_1 + \omega < \omega_1 \cdot \omega < \omega_1^{\omega}$.

Proof. (a) $\omega + \omega_1 = \sup\{\omega + \alpha : \alpha < \omega_1\}$, but each such $\omega + \alpha$ is countable, so $\sup\{\omega + \alpha : \alpha < \omega_1\} = \omega_1$. The proofs of (b) and (c) are similar. For (d), $\omega_1 < \omega_1 + 1 < \omega_1 + \omega < \omega_1 + (\omega + 1) < \omega_1 + \omega_1 = \omega_1 \cdot 2 < \omega_1 \cdot \omega < \omega_1 \cdot (\omega + 1) < \omega_1 \cdot \omega_1 = \omega_1^2 < \omega_1^{\omega}$.

As in arithmetic on the natural numbers, ordinal arithmetic has the concepts of division and remainder.

Theorem 23. Let $0 < \alpha < \beta$ be ordinals. Then there are ordinals δ, γ with $\gamma < \alpha$ and $\beta = \alpha \cdot \delta + \gamma$.

Proof. Let δ be the sup of $\{\rho : \alpha \cdot \rho \leq \beta\}$. By definition of δ, $\alpha \cdot \delta \leq \beta$ and $\alpha \cdot (\delta + 1) > \beta$. Let γ be the order type of $\{\eta : \alpha \cdot \delta \leq \eta < \beta\}$. Then $\alpha \cdot \delta + \gamma = \beta$ and $\gamma < \alpha$ since $\beta < \alpha \cdot (\delta + 1)$.

Ordinal arithmetic does not have a right-cancellation law. As the next theorem shows, counterexamples to right-cancellation are common.

Theorem 24. For all ordinals β and all nonzero limit ordinals α, $\alpha + \beta = \beta$ iff $\beta \geq \alpha \cdot \omega$.

Proof. If $\alpha + \beta = \beta$, then $\beta \geq \alpha$, so there are δ, γ with $\beta = \alpha \cdot \delta + \gamma$, $\gamma < \alpha$. Then

$$\alpha + \beta = \alpha + (\alpha \cdot \delta) + \gamma = \alpha(1 + \delta) + \gamma \begin{cases} = \alpha\delta + \gamma & \text{if } \delta \geq \omega \\ > \alpha\delta + \gamma & \text{if } \delta < \omega. \end{cases}$$

Theorem 24 says the operation $\alpha +$ has many fixed points. The operation $\alpha \cdot$ also has many fixed points (see corollary 29). The operation $+ \alpha$ (if $\alpha > 0$) and $\cdot \alpha$ (if $\alpha > 1$) have none.

Theorem 25. For all ordinals α, β, γ, if $\beta < \gamma$, then $\alpha + \beta < \alpha + \gamma$, and if $\beta < \gamma$ and $\alpha > 0$, then $\alpha \cdot \beta < \alpha \cdot \gamma$.

Proof. Immediate from definitions 13 and 14.

It will turn out that, as a corollary to theorem 28, there will be many pairs α, β with $\alpha + \beta = \alpha \cdot \beta$; as a corollary to theorem 25, there are no β satisfying $\beta + \beta = \beta \cdot \beta$ if $\beta > 2$.

For completeness, we define infinite ordinal operations:

Definition 26. For each ordinal β, and each set of ordinals $\{\alpha_\gamma : \gamma < \beta\}$.

    (a) $\sum_{\gamma < \beta} \alpha_\gamma = \sup\{\sum_{\gamma < \delta} \alpha_\gamma : \delta < \beta\}$ if β is a limit; if $\beta = \eta + 1$, then $\sum_{\gamma < \beta} \alpha_\gamma = \sum_{\gamma < \eta} \alpha_\gamma + \alpha_\eta$.

    (b) $\prod_{\gamma < \beta} \alpha_\gamma = \sup\{\prod_{\gamma < \delta} \alpha_\gamma : \delta < \beta\}$ if β is a limit; if $\beta = \eta + 1$, then $\prod_{\gamma < \beta} \alpha_\gamma = (\prod_{\gamma < \eta}) \alpha_\gamma \cdot \alpha_\eta$.

For example, in ordinal arithmetic, $0 + 1 + 2 + \cdots + \omega = \omega + \omega$; $0 \cdot 1 \cdot 2 \cdot \cdots \cdot \omega = \omega \cdot \omega$.

The operations of definition 26 are explored in the exercises.

We close this section with a theorem about continuity and fixed points on the ordinals which establishes many of the fixed-point theorems for ordinal arithmetic.

Definition 27. A class function f from ON to ON is said to be continuous iff it is nondecreasing and, for all limit α, $f(\alpha) = \sup\{f(\beta) : \beta < \alpha\}$.

For example, every constant function is continuous, and the function $f(\alpha) = \beta + \alpha$ is continuous, for each fixed β, as are the functions $g(\alpha) = \beta \cdot \alpha$ and $h(\alpha) = \beta^\alpha$.

Theorem 28. A continuous strictly increasing class function defined on all the ordinals has arbitrarily high fixed points; i.e., for every α there is some $\beta > \alpha$ with $f(\beta) = \beta$.

    Proof. Let $\alpha_0 = f(\alpha)$ and define $\alpha_{n+1} = f(\alpha_n)$ for all finite n. Let $\alpha^* = \sup\{\alpha_n : n \in \omega\}$. By continuity, $f(\alpha^*) = \alpha^*$.

Corollary 29

    (a) Let α be a nonzero limit ordinal. Then there are arbitrary high β for which $\alpha \cdot \beta = \beta$.

(b) For every nonzero limit ordinal α, there are arbitrary high β with $\alpha + \beta = \alpha \cdot \beta$.

Proof

(a) The function $f(\beta) = \alpha \cdot \beta$ is continuous, strictly increasing.
(b) Let β be a fixed point for the function $f(\beta) = \alpha \cdot \beta$ where $\beta \geq \alpha \cdot \omega$. Then by theorem 24, $\beta = \alpha + \beta$.

SECTION 5.3. CARDINAL ARITHMETIC

In this section we study cardinal arithmetic in the presence of the axiom of choice. (Since without the axiom of choice there is almost nothing to say beyond the Schröder–Bernstein theorem, we will, in this section and for the rest of the book, automatically assume the axiom of choice unless otherwise noted, thus reversing our previous convention of always mentioning explicitly the use of choice.)

We introduce the following conventions: Lowercase Greek letters in the first part of the alphabet (α, β, γ, δ, and so on) represent ordinals, and arithmetic operations involving them (e.g., $\alpha + \beta$) are always ordinal operations; later lowercase Greek letters (e.g., κ, λ, ρ) represent cardinals, and operations involving only them (e.g., $\kappa + \lambda$) are cardinal operations unless noted otherwise. Mixed operations (e.g., $\lambda + \alpha$) are ordinal operations as are all $\lambda + n$ where n is finite, unless otherwise noted.

Let us define cardinal addition and multiplication. If you have three rings on one hand and four rings on the other, you have seven rings all together. That is how we define cardinal addition: the size of the union of two nonoverlapping sets.

If a matrix has 35 rows and 20 columns, it has 700 entries. That is how we define cardinal multiplication: as the size of a product.

Definition 30

(a) Let X and Y be sets. We define $|X| + |Y| = |(X \times \{0\}) \cup (Y \times \{1\})|$.
(b) Let X, Y be sets. $|X| \cdot |Y| = |X \times Y|$.

Note that this definition does not depend on the axiom of choice. Let us give the version whose universality depends on choice.

Definition 31. Let κ, λ be cardinals. We define $\kappa + \lambda = \rho$ iff there are disjoint sets A, B with $|A| = \kappa$, $|B| = \lambda$, and $|A \cup B| = \rho$; and $\kappa \cdot \lambda = |\kappa \times \lambda|$.

The reader is expected to show that definition 31 is consistent with definition 30.

We already defined one special case of cardinal exponentiation when we defined 2^κ as $|^\kappa 2|$. Note that this is consistent with ordinary exponentiation on the integers. Also consistent with exponentiation on the integers is

Definition 32. For any sets $X, Y, |Y|^{|X|} = |^X Y|$. In particular, if κ, λ are cardinals, $\lambda^\kappa = |^\kappa\lambda|$.

Note that definition 32 again does not need the axiom of choice to apply to all sets.

To prove the main theorem about cardinal addition and multiplication it will be convenient to define two classes of ordinals.

Definition 33. An ordinal is even iff it has the form $\alpha + 2n$ where $n \in \omega$ (here addition means ordinal addition) and α is a limit ordinal. An ordinal which is not even is odd.

Lemma 34. Let κ be an infinite cardinal. Then $\kappa = |\{\alpha < \kappa: \alpha \text{ is even}\}| = |\{\alpha < \kappa: \alpha \text{ is odd}\}|$.

　　Proof. Let $f(\alpha + 2n) = \alpha + n$ where α is a limit. Then f is 1–1 from the set of even ordinals below κ onto κ. And letting $g(\alpha + 2n + 1) = \alpha + 2n$ we see that the set of even ordinals below κ has the same cardinality as the set of odd ordinals below κ.

An immediate consequence of lemma 34 is that, for all infinite cardinals κ, $\kappa + \kappa = \kappa$.

Thus, while the definitions of infinite cardinal operations parallel their finite counterparts, their salient arithmetical properties can be quite different. It is easy to prove that cardinal addition and multiplication are commutative and associative and that the usual distributive law holds. But, unlike finite arithmetic, infinite cardinal operations are idempotent.

Theorem 35. Let κ, λ be infinite cardinals. Then

(a) (Idempotence) $\kappa + \kappa = \kappa = \kappa \cdot \kappa$.

(b) $\kappa + \lambda = \sup\{\kappa, \lambda\}$.

(c) $\kappa \cdot \lambda = \sup\{\kappa, \lambda\}$.

Proof. For (a): We have already shown that $\kappa + \kappa = \kappa$. We must show that $\kappa \cdot \kappa = \kappa$. We do this by induction on infinite cardinals. Induction on cardinals reduces to ordinal induction as follows: Assuming a hypothesis is true for all infinite cardinals $\rho < \kappa = \omega_\alpha$ is assuming that it is true for all ω_β, $\beta < \alpha$. So the induction is on the ordinal indices of the cardinals.

To get started, note that the proof that the rationals are countable is essentially the proof that $\omega \cdot \omega = \omega$.

Now suppose for all infinite cardinals $\rho < \kappa$, $\rho \cdot \rho = \rho$. Well-order the set $\kappa \times \kappa$ as follows: $(\alpha, \beta) \leq (\gamma, \delta)$ iff the ordinal sum $\alpha + \beta <$ the ordinal sum $\gamma + \delta$; or the ordinal sum $\alpha + \beta =$ the ordinal sum $\gamma + \delta$ and $\alpha < \gamma$. This is easily seen to be a linear ordering. It is a well-ordering: For any subset A of $\kappa \times \kappa$ there is a least ordinal sum α of pairs in A, and the set of pairs in A which sum to α has a pair with a least first element. We are done if we can show that the order type of this well-ordering is exactly κ.

Let $B_\alpha = \{(\gamma, \beta) \in \kappa \times \kappa : \gamma + \beta = \alpha\}$, for each $\alpha < \kappa$. Then $\kappa \times \kappa = \bigcup_{\alpha < \kappa} B_\alpha$, and by induction hypothesis each B_α has size at most $|\alpha|$ (because each element of B_α is in $(\alpha + 1) \times (\alpha + 1)$). Let $C_\beta = \bigcup_{\alpha < \beta} B_\alpha$ for each $\beta < \kappa$. Note that $C_\beta \subset \beta \times \beta$. By the induction hypothesis each $|\beta| \leq |C_\beta| < \kappa$, so $\kappa \times \kappa$ has, under this ordering, length at least κ. Each C_β is an initial segment of $\kappa \times \kappa$, so $\kappa \times \kappa$ has, under this ordering, length at most κ. Thus the length is exactly κ, and we are done.

For (b): Without loss of generality suppose $\kappa \leq \lambda$. Then $\lambda \leq \kappa + \lambda \leq \lambda + \lambda$ which, as we already know, equals λ.

For (c): Again, without loss of generality suppose $\kappa \leq \lambda$. Again, $\lambda \leq \kappa \cdot \lambda \leq \lambda \cdot \lambda$. But, by (a), $\lambda \cdot \lambda = \lambda$ for all cardinals λ, and we are done.

Note that the statement $\kappa^2 = \kappa$ is equivalent to the statement that the arbitrary union of κ many sets, each of size κ, has size κ.

Theorem 36

(a) For every infinite cardinal κ, $2^\kappa = \kappa^\kappa$.

(b) If κ, λ are cardinals, λ infinite, and $2 \leq \kappa \leq \lambda$, then $\kappa^\lambda = \lambda^\lambda$.

(c) If κ, λ, ρ are cardinals, then $(\kappa^\lambda)^\rho = \kappa^{\lambda \cdot \rho}$.

Proof. For (a): A function in $^\kappa\kappa$ is a subset of $\kappa \times \kappa$, and we already know that $|\kappa \times \kappa| = \kappa$, from theorem 35. So there is a 1–1 correspondence from $^\kappa\kappa$ to $\mathcal{P}(\kappa)$; hence $|^\kappa\kappa| \leq 2^\kappa \leq \kappa^\kappa$, and we are done.

Note that (b) is an immediate corollary of (a).

For (c), there is a 1–1 mapping Φ from functions from ρ into κ^λ to functions from $\rho \times \lambda$ into κ given as follows: If $f: \rho \to \kappa^\lambda$, let $\Phi(f)$ be defined by $\Phi(f)(\alpha, \beta) = f(\alpha)(\beta)$.

Notice that theorem 36 does not tell us exactly what any κ^λ is. Theorem 35 told us exactly which cardinal was the sum or product of two infinite cardinals. We simply do not know this for exponentiation. 2^ω could be nearly anything in a very precise sense: If κ is an infinite cardinal of uncountable cofinality (defined in the next section), then it is consistent with the axioms of set theory that $2^\omega = \kappa$. A little more can and will be said in the next section, but not much more. Most questions involving exponentiation are undecidable. When one of them is discovered not to be, it is a major breakthrough (see the discussion of the singular cardinals problem in section 5.5).

Although we cannot know what 2^ω really is (whatever that means), we can hypothesize what it might be. The continuum hypothesis (abbreviated CH) is the hypothesis that $2^\omega = \omega_1$; the general continuum hypothesis (abbreviated GCH) is that, for every infinite cardinal κ, $2^\kappa = \kappa^+$. Cantor believed that CH was true and spent many fruitless years trying to prove it so. In 1937 Gödel proved that GCH is consistent by showing that it holds in a particular model of set theory known as the constructible universe (see chapter 6). In 1963 Cohen proved that CH does not hold in all models of set theory, via the technique of forcing (invented by Cohen to show the consistency of the failure of CH and the consistency of the failure of AC; forcing has become the main method of showing statements consistent with ZF or with ZFC).

SECTION 5.4. COFINALITY

Our previous classifications of infinite cardinals—into countable and uncountable cardinals and into successor and limit cardinals—are far too coarse: The former even lumps all infinite cardinals except ω into the same class. Here we develop a far more useful classification, based on the concept of cofinality.

Recall that a set A is said to be cofinal in a linear order X iff $A \subset X$ and for every $x \in X$ there is $a \in A$ with $x \le a$. In particular, if $A \subset \alpha$ and α is a limit ordinal, then A is cofinal in α iff $\alpha = \bigcup A$.

Definition 37. An ordinal α is cofinal in an ordinal β iff there is a 1–1 increasing function from α to β whose range is cofinal in β.

For example, an infinite ordinal α is cofinal in $\beta + \alpha$, and in $\beta \cdot \alpha$ and β^α if α is a limit (the arithmetic here is ordinal arithmetic), and in the cardinal ω_α if α is a limit. Each $\alpha > 0$ is cofinal in α.

Note that if α is cofinal in β then $\alpha \le \beta$ and that if α is cofinal in β which is cofinal in γ then α is cofinal in γ.

Definition 38. The cofinality of an ordinal β is defined to be the least α with α cofinal in β. We write $\alpha = \mathrm{cf}(\beta)$.

If β is a successor ordinal, then $\mathrm{cf}(\beta) = 1$.

What sort of ordinals can be cofinalities?

Theorem 39

(a) An infinite ordinal of size κ has cofinality at most κ.

(b) If $\alpha = \mathrm{cf}(\beta)$ for some β, then α is a cardinal.

Proof. For (a): Suppose $|\alpha| = \kappa$ and $\rho = \mathrm{cf}(\alpha)$. Let $f: \kappa \to \alpha$ be 1–1 and onto, and define g with dom $g \subset \kappa$, range g cofinal in α by $g(\beta) = \inf\{f(\gamma): \gamma \ge \beta$ and $f(\gamma) > g(\delta)$ for all $\delta < \beta\}$. For some $\beta \le \kappa$, dom $g = \beta$. Note that g is a strictly increasing function on $\beta \le \kappa$ whose range is cofinal in α, so $\beta \ge \rho$.

For (b): If $|\alpha| = \kappa < \alpha = \mathrm{cf}(\beta)$, then by (a) $\mathrm{cf}(\alpha) \le \kappa$. Hence some $\delta \le \kappa$ is cofinal in α which is cofinal in β, so α is not the least ordinal cofinal in β, a contradiction.

On the other hand, not every cardinal is a cofinality, witness $\aleph_\omega$.

The important bifurcation of cardinals is given by

Definition 40. A cardinal κ is regular iff $\kappa = \mathrm{cf}(\kappa)$. Otherwise κ is singular.

For example, $\aleph_\omega$ is singular, but ω_1 is regular. The latter follows from

Theorem 41. For every cardinal κ, κ^+ is regular (recall that κ^+ is the first cardinal strictly greater than κ).

Proof. By theorem 39, $\mathrm{cf}(\kappa^+) = \kappa^+$ or $\mathrm{cf}(\kappa^+) \le \kappa$. If the former, we are done, so suppose the latter. Then letting $\lambda = \mathrm{cf}(\kappa^+)$, we have an increasing cofinal 1–1 map $f: \lambda \to \kappa^+$, where, if $\alpha < \lambda$, $A_\alpha = \{\gamma < \kappa^+: \gamma < f(\alpha)\}$, then κ^+ is the union of the A_α's. But then κ^+ is the union of at most κ sets each of which has size at most κ, so $|\kappa^+| = \kappa$, which is a contradiction.

We have shown that every infinite successor cardinal is regular. Are there any regular limit cardinals? Clearly ω is one. In the next chapter we will show that the existence of an uncountable regular limit cardinal implies the consistency of ZF. Hence, by Gödel's incompleteness theorems, ZF (and in fact ZFC) cannot prove the existence of a regular uncountable limit cardinal. Such a cardinal is called weakly inaccessible. (Strongly inaccessible cardinals will be defined in the next chapter.) On the other hand, the existence of an uncountable regular limit cardinal seems like little enough to ask of the universe, and inasmuch as set theorists can be said to believe in the existence of the mathematical universe (whatever "existence" means in this context) they can probably be said to believe in the existence of weakly inaccessible cardinals. Certainly theorems which begin "if there is a weakly inaccessible cardinal, then..." are regarded as innocuous. But, on the other hand, when a theorem does begin that way, an effort is always made to check that the hypothesis is necessary.

SECTION 5.5. INFINITE OPERATIONS AND MORE EXPONENTIATION

Generalizing section 5.3, we define infinite cardinal sum and infinite cardinal product as follows:

Definition 42. Let I be an index set, κ_i a cardinal for each $i \in I$. We define $\sum_{i \in I} \kappa_i = |\bigcup_{i \in I} A_i|$ where each $|A_i| = \kappa_i$ and the A_i's are pairwise disjoint. In an abuse of notation we define $\prod_{i \in I} \kappa_i = |\prod_{i \in I} \kappa_i|$ (here the first $\prod$ means cardinal product, the second means set-theoretic product).

The connection between infinite sum and infinite product is given by König's theorem:

Theorem 43. Suppose $I \neq \emptyset$ and $\kappa_i < \lambda_i$ for all $i \in I$. Then $\sum_{i \in I} \kappa_i < \prod_{i \in I} \lambda_i$.

Proof. For each $i \in I$ let A_i be such that $|A_i| = \kappa_i$ where the A_i's are pairwise disjoint, and let f_i be a 1-1 map from A_i into $\lambda_i - \{0\}$. For $x \in A_i$ let $f_x(i) = f_i(x)$; $f_x(j) = 0$ if $j \neq i$. The map $F(x) = f_x$ shows that $\sum_{i \in I} \kappa_i \leq \prod_{i \in I} \lambda_i$.

To show that $\sum_{i \in I} \kappa \neq \prod_{i \in I} \lambda_i$, we let F be an arbitrary map from $\bigcup_{i \in I} (\kappa_i \times \{i\})$ into $\prod_{i \in I} \lambda_i$ and show that F is not onto. Let $g_{\alpha,i}$ denote $F(\alpha, i)$, and, for each $i \in I$ let $K_i = \{g_{\alpha,i}(i): \alpha < \kappa_i\}$. Since each $|K_i| \leq$

$\kappa_i < \lambda_i$, there is some $\gamma_i \in \lambda_i - K_i$. Let $f(i) = \gamma_i$ for all $i \in I$. Then $f \in$ range F iff $f = g_{\alpha,i}$, for some α, i, iff $f(i) \in K_i$ for some i; hence $f \notin$ range F.

With the concept of cofinality and theorem 43 at hand we can get sharper theorems about exponentiation. The basic elementary theorem is

Theorem 44. For each infinite cardinal κ

(a) $\kappa^{\text{cf}(\kappa)} > \kappa$, and
(b) cf $2^\kappa > \kappa$.

Proof. For (a): Let f be a 1–1 increasing cofinal map from cf κ to κ. Then $\kappa = |\bigcup_{\alpha<\text{cf}(\kappa)} f(\alpha)| \leq \sum_{\alpha<\text{cf}(\kappa)} |f(\alpha)| < \kappa^{\text{cf}\,\kappa}$ by König's lemma.
For (b): Let $\lambda = $ cf 2^κ. Then $(2^\kappa)^\lambda = 2^\kappa$ if $\lambda \leq \kappa$, contradicting (a).

Note that, since $\text{cf}(\lambda) \leq \lambda$ for all λ, theorem 44(b) implies the infinitary case of Cantor's theorem that $\kappa < 2^\kappa$.

Theorem 44 nearly ends what can be said about exponentiation without getting into consistency results. (For example, we cannot prove that $2^\kappa \geq 2^\lambda$ implies $\kappa \geq \lambda$.) The exceptions are the remarkable results due to Silver and to Galvin and Hajnal on the singular cardinals problem, and Bukovsky's theorem[†] (proved independently by Hechler) that κ^κ is determined by $\kappa^{\text{cf}(\kappa)}$. To put these results in context, and to explain the situation for regular cardinals, we first state

Theorem 45 (Easton). Let F be a nondecreasing function from the class of infinite regular cardinals to the class of cardinals where cf $F(\kappa) > \kappa$ for all κ. (Note that F is a class and that $F(\kappa)$ need not itself be regular.) Then if ZFC is consistent, so is ZFC + "for all regular κ, $2^\kappa = F(\kappa)$."

Easton's theorem is proved via forcing, the technique created by Cohen in 1963 to construct new models of set theory from old models, and is a complicated generalization of Cohen's original proof that 2^ω need not be ω_1. It completely settles the question of what the possibilities are for 2^κ if κ is regular, saying that the constraint of theorem 44(b) is the only constraint possible. The proof of Easton's theorem, as with all forcing proofs, is beyond the scope of this book.

[†]A piece of the Bukovsky–Hechler theorem appears as exercise 28. For the full theorem, see Jech's *Set Theory*.

After Easton's theorem the question became "what about singular cardinals? Does the same sort of freedom exist there?" The surprising answer, due to Silver, was "not if the cofinality is uncountable."

Theorem 46 (Silver). Suppose $\aleph_\alpha$ is a singular cardinal of uncountable cofinality and, for some fixed $\gamma < \mathrm{cf}(\alpha)$ and every ordinal $\beta < \alpha$, $2^{\aleph_\beta} = \aleph_{\beta+\gamma}$. Then $2^{\aleph_\alpha} = \aleph_{\alpha+\gamma}$.

The version of Silver's theorem we have given is not the strongest possible (for example, the hypothesis need not hold for every $\beta < \alpha$, just for a large enough set of them), but it is the strongest version which is easy to state. Note that as an immediate corollary we have that, if κ is a singular cardinal with uncountable cofinality and $2^\lambda = \lambda^+$ for all cardinals $\lambda < \kappa$, then $2^\kappa = \kappa^+$. It is this consequence which is most widely known, and which made Silver's theorem surprising, since Prikry and Silver had previously shown that, modulo the consistency of a cardinal stronger than a measurable cardinal (see chapter 7), there could be a singular cardinal κ of countable cofinality where $2^\lambda = \lambda^+$ for each infinite $\lambda < \kappa$ but $2^\kappa = \kappa^{++}$. ("There could be" means that it happens in some models of set theory, not that it happens in all of them.)

Then Baumgartner, Prikry, and Jensen independently gave a purely elementary (in the sense that it only uses combinatorial set theory and no techniques from mathematical logic) proof of Silver's theorem which Galvin and Hajnal generalized in still other directions. Galvin and Hajnal's proof, while elementary in the sense that it uses no model theory, is quite a bit more advanced than the level of this book. It has as a consequence

Theorem 47 (Galvin, Hajnal). If κ is a singular cardinal of uncountable cofinality and if $2^\lambda < \kappa$ for all $\lambda < \kappa$ and if $\kappa < \aleph_\kappa$, then $2^\kappa < \aleph_\kappa$.

The reader interested in seeing their proof is directed to either Jech's *Set Theory*, which gives the entire proof, or to the text by van Dalen, Doets, and Swart (listed in the bibliography) which gives a special case of their proof.

Further work on the singular cardinals problem for uncountable cofinality has been done by Jensen, who, in a remarkable result which has inspired analogues in other contexts, shows that the negation of various large cardinal axioms also affect the possibilities for 2^κ when κ is singular. His work will be described briefly in chapter 6. Shelah has obtained exponentiation bounds in the case of countable cofinality.

SECTION 5.6. COUNTING

In this section we compute the size of various mathematical objects.

Example 48. $\mathbb{R}$. By section 4.4, every real corresponds to a set of rationals. So $|\mathbb{R}| \leq \omega^\omega = 2^\omega$. In the proof of corollary 9(b) we showed that $|\mathbb{R}| \geq 2^\omega$. So $|\mathbb{R}| = 2^\omega$.

Example 49. *Polynomials.* How many polynomials are there in one variable x with coefficients in $\mathbb{R}$? Such a polynomial has the form $\sum_{i \leq n} r_i x^i$, where n is finite and each $r_i \in \mathbb{R}$. So there are exactly as many polynomials as there are finite tuples with elements in $\mathbb{R}$. Thus there are $|\bigcup_{n \in \omega} \mathbb{R}^n| = \omega \cdot 2^\omega = 2^\omega$ polynomials with coefficients in $\mathbb{R}$.

Similarly, there are $\omega \cdot \omega = \omega$ polynomials in one variable x with coefficients in $\mathbb{Q}$.

Example 50. *The number of open sets of reals.* Recall that an open set of reals is a union of open intervals with rational endpoints. Let $\mathscr{J}$ be the set of intervals with rational endpoints. If u is open, we let $f(u) = \{J \in \mathscr{J} : J \subset u\}$. The function f is 1-1 and $\mathscr{J}$ is countable, so there are at most 2^ω many open sets of reals. That there are exactly 2^ω open sets of reals follows from the fact that, for every real r, $\mathbb{R} - \{r\}$ is open.

Example 51. *The number of irrationals.* There are countably many rationals, and 2^ω many reals, so there must be 2^ω many irrationals.

Example 52. *The number of continuous functions from $\mathbb{R}$ to $\mathbb{R}$.* There are $2^{(2^\omega)}$ functions from $\mathbb{R}$ to $\mathbb{R}$. But there are only 2^ω continuous functions from $\mathbb{R}$ to $\mathbb{R}$, as follows:

Every continuous function from $\mathbb{R}$ to $\mathbb{R}$ is determined by its values on $\mathbb{Q}$, so there are at most $^{|\mathbb{Q}|}|\mathbb{R}| = (2^\omega)^\omega = 2^\omega$ many such functions. There are at least that many: For each real r the constant function $f(x) = r$ is continuous.

Example 53. *The number of 1-1 functions from ω into ω_1.* First we show that there are ω_1^ω 1-1 functions from ω into ω_1. Surely there are at most that many. It suffices to find a 1-1 function G from $^\omega \omega_1$ into the set of 1-1 functions from ω into $\omega \times \omega_1$. If $f : \omega \to \omega_1$, define $G(f) = \{(n, (n, \alpha)) : f(n) = \alpha\}$. G is clearly a 1-1 function, as is $G(f)$.

Now note that, since $\omega < \omega_1 \leq 2^\omega$, $2^\omega \leq \omega_1^\omega \leq (2^\omega)^\omega = 2^\omega$. Thus there are 2^ω 1-1 functions from ω into ω_1.

Example 54. *Subsets of κ of size κ.* Clearly 2^κ is an upper bound. If A is a subset of κ, let $f(A) = \bigcup_{\alpha \in A} \{\alpha\} \times \kappa$. Then f is a 1-1 function from $\mathscr{P}(\kappa)$ into $\mathscr{P}(\kappa^2)$ where each $f(A)$ has size κ. Since $\kappa = |\kappa^2|$, the number of subset of κ of size κ is at least 2^κ; hence it equals 2^κ.

Example 55. *Well-orders on κ.* By definition, there are exactly κ^+ ordinals of size κ. We show that there are 2^κ ways in which to well-order a set of size κ. (Note that by theorem 45 2^κ need not equal κ^+.)

2^κ is an upper bound, since every well-ordering of κ is a relation on κ, i.e., a set of ordered pairs in κ^2. Let A be a subset of κ of size κ. By example 54 there are 2^κ of these. Since κ is a cardinal, A is well-ordered by $\in$ in order type κ. Let $\leq_A$ be any well-ordering of κ in which A under $\in$ is an initial segment. Since the $\leq_A$'s are distinct, we are done.

EXERCISES FOR CHAPTER 5

1. Show by the methods of definition 1 that the following are countable sets: $\omega \times \omega$; $\omega \times \omega \times \omega$; the set of polynomials in one variable with integer coefficients.

2. Show by the methods of section 5.1 that each of the following has the same cardinality as $\mathbb{R}$: $\mathbb{R}^2$, $\mathbb{R}^n$ for each finite n, the set of points on the surface of the unit sphere.

3. Using exercise 2 and the Schröder–Bernstein theorem, show that $|\mathbb{R}| = |\text{unit ball}|$.

4. An exercise on the proof of the Schröder–Bernstein theorem: Let $A = \{0, 1\} \times \omega$, $B = \omega$, $f: A \to B$ where $f(0, n) = 4n$, $f(1, n) = 4n + 3$, $g: B \to A$ where $g(n) = (0, n)$.
   (a) What is each A_n? B_n?
   (b) What is $A^\dagger$? $B^\dagger$?

Note: In exercises 5 through 7 you may only use the methods of section 5.1.

5. (a) Show, assuming AC, that every infinite set has a countably infinite subset.
   (b) Show, assuming AC, that if X is infinite and Y countable then $|X| = |X \cup Y|$.

6. Assuming AC, show that $|\mathbb{R}| \geq \omega_1$.

7. (a) Show that if $|x| > 1$ then $|x| < |^x x|$.
   (b) Assuming AC, show that $|^R R| \geq \omega_2$.

Note: In Exercises 8 through 15 all arithmetic is ordinal arithmetic.

8. Find a set of reals of order type $\omega + \omega + \omega + \omega + \omega$.

9. Show that $\gamma \cdot \beta \cdot \alpha$ is order-isomorphic to $\alpha \times \beta \times \gamma$ under the lexicographic ordering.

10. Show that example 18(c) works.

11. Suppose we have embedded α into $\mathbb{R}$ via f, where $f(\beta) = x_\beta$ for each $\beta < \alpha$, and each x_β lies strictly between 0 and 1. Now let us embed $\alpha \cdot \omega$ as follows: if $i_1 i_2 i_3 \ldots$ is the decimal expansion of x_β, we let $x_{\beta,n}$ be the number whose decimal expansion is $.\underbrace{1 \ldots 1}_{n \text{ ones}} 0 i_1 i_2 i_3 \ldots$. Show that $\{x_{\beta,n} : \beta < \alpha, n < \omega\}$ has order type $\alpha \cdot \omega$.

12. In this exercise we will find a subset of $\mathbb{R}$ of order type ω^ω. For each $(k_1, k_2, \ldots, k_n) \in \omega^n$, let $x_{k_1 k_2 \ldots k_n}$ be the real whose decimal expansion is

$$.\underbrace{1 \ldots 1}_{n \text{ ones}} 0 \underbrace{1 \ldots 1}_{k_1 \text{ ones}} 0 \underbrace{1 \ldots 1}_{k_2 \text{ ones}} 0 \ldots 10 \ldots 01 \underbrace{\ldots 1000}_{k_n \text{ ones}} \ldots$$

   (a) Show that if $\mathbf{k}, \mathbf{j} \in \omega^n$ and $\mathbf{k} <_L \mathbf{j}$, then $x_{\mathbf{k}} < x_{\mathbf{j}}$. Conclude that $\{x_{\mathbf{k}} : \mathbf{k} \in \omega^n\}$ has order type ω^n.
   (b) Show that if $n < m$, $\mathbf{k} \in \omega^n$, and $\mathbf{j} \in \omega^m$, then $x_{\mathbf{k}} < x_{\mathbf{j}}$.
   (c) Show that $\{x_{\mathbf{k}} : \text{for some } n, \mathbf{k} \in \omega^n\}$ has order type ω^ω.

13. List the following ordinals in nondecreasing order. State which ordinals are in fact equal to each other: ω^{ω_1}; 3^ω; $\omega_1{}^\omega$; ω; ω_1; ω^3; $\omega \cdot 3$; $\omega \cdot \omega_1$; $\omega_1 \cdot \omega$.

14. Now do exercise 10 replacing each ω_1 by ω_{17}, ω_2 by ω_{20}, ω_3 by ω_{21} (e.g., $\omega_1 \cdot \omega$ becomes $\omega_{17} \cdot \omega_{20}$).

15. List the following ordinals in nondecreasing order. State which ordinals are in fact equal to each other (all arithmetic is ordinal arithmetic): $\sum_{i<\omega} i$, $\sum_{i<\omega} \omega \cdot i$, ω, $\prod_{i \leq \omega} i$, $\prod_{i<\omega} \omega \cdot i$, $\omega \cdot \omega$, ω^ω, $\sum_{i<\omega} \omega^i$, $\prod_{i<\omega} \omega^i$.

Note: In exercises 16 through 33, all arithmetic is cardinal arithmetic. You may assume AC.

16. (a) Show that $\kappa + \kappa = 2 \cdot \kappa$.
    (b) Show that $\kappa^2 = \kappa \cdot \kappa$.

17. (a) Show that cardinal addition and multiplication are associative.

(b) Show that cardinal addition and multiplication are commutative.

(c) Show that $\kappa \cdot (\lambda + \rho) = \kappa \cdot \lambda + \kappa \cdot \rho$.

18. (a) Show that no function from ω_1 to ω_2 is onto.

(b) (Hausdorff) Show that $\omega_2^{\omega_1} = 2^{\omega_1}$.

19. Arrange the following cardinals in nondecreasing order. State which cardinals are in fact equal to each other:

$$\omega^{\omega_1}; \; 3^\omega; \; \omega_1^\omega; \; \omega; \; \omega^3; \; \omega \cdot 3; \; \omega \cdot \omega_1; \; \omega_1 \cdot \omega$$

20. Show that $\mathrm{cf}(\mathrm{cf}(\alpha)) = \mathrm{cf}(\alpha)$ for all ordinals α.

21. Assume GCH. Show that if κ is weakly inaccessible and $\lambda < \kappa$ then $2^\lambda < \kappa$.

22. (a) Show that every singular cardinal κ is the union of fewer than κ sets each of which has size $< \kappa$.

(b) Show that no regular cardinal κ is the union of fewer than κ sets each of which has size $< \kappa$.

23. (a) Show that if κ is regular and $\lambda < \kappa$ then $\kappa^\lambda = \kappa \cdot \sup\{|\rho^\lambda|: \rho < \kappa\}$.

(b) Show that if κ is weakly inaccessible and $\lambda < \kappa$ then $\kappa^\lambda = \sup\{|\rho^\lambda|: \rho < \kappa\}$.

(c) Assuming GCH, show that if κ is regular and $\lambda < \kappa$ then $\kappa^\lambda = \kappa$.

24. Show that the definitions of infinite cardinal sum and product are independent of the arrangement of the sets; i.e., if $f: I \to J$ is 1–1 and onto, then $\sum_{i \in I} \kappa_i = \sum_{i \in I} \kappa_{f(i)}$ and $\prod_{i \in I} \kappa_i = \prod_{i \in I} \kappa_{f(i)}$.

25. Show that each $\kappa \cdot \lambda = \sum_{\gamma < \lambda} \kappa_\gamma$ and each $\kappa^\lambda = \prod_{\gamma < \lambda} \kappa_\gamma$ where each $\kappa_\gamma = \kappa$.

26. (a) Show that each $\sum_{i \in I} \kappa_i = |\sup\{\kappa_i: i \in I\}| + |I|$, where each $\kappa_i \geq \omega$.

(b) Give an example of some $\prod_{i \in I} \kappa_i$ which does not equal $|\sup\{\kappa_i: i \in I\}| + |I|$, where each $\kappa_i \geq \omega$.

27. Use König's theorem (theorem 43) to give a one-line proof that each $\kappa < 2^\kappa$.

28. Show that if $2^\lambda < \kappa$ for every $\lambda < \kappa$ then $2^\kappa = \kappa^{\mathrm{cf}(\kappa)}$.

29. Which of the following statements are true, which are false, and which are independent?

(a) $2^\omega = \omega_1$.

(b) $2^\omega = \aleph_\omega$.

(c) $2^\omega < 2^{\omega_1}$.

(d) $2^\omega \leq 2^{\omega_1}$.

(e) $2^\omega = \aleph_{\omega_1}$.

(f) $2^\omega = \omega_{17}$.

(g) Let $\kappa = \aleph_{\omega_1}$. If $2^\lambda = \lambda^{++}$ for $\lambda < \kappa$, then $2^\kappa = \kappa^{++}$.

(h) Let $\kappa = \aleph_{\omega_1}$. $2^\lambda < \kappa$ for $\lambda < \kappa$ and $2^\kappa = \aleph_{\kappa^+}$.

30. Recall the formal definition of a linear algebra with basis B over $\mathbb{R}$ as the set of terms $\sum_{i \leq n} r_i b_i$ where each $r_i \in \mathbb{R}$, each $b_i \in B$, and n is finite. If $|B| = \kappa \geq \omega$, how big is the linear algebra with basis B over $\mathbb{R}$?

31. A G_δ set is a countable intersection of open sets. How many G_δ sets are there in $\mathbb{R}$? In each $\mathbb{R}^n$ (n finite)?

32. (a) Let F be the filter on $\kappa \geq \omega$ defined by $a \in F$ iff $\kappa - a$ is finite. How big is F?

(b) Let G be the filter on $\kappa \geq \omega$ defined by $a \in F$ iff $|\kappa - a| < \kappa$. How big is G?

33. How many 1–1 functions are there from ω into ω_2? From ω_1 into ω_2?

6

TWO MODELS OF
SET THEORY

INTRODUCTION

In this section we develop two models of set theory, V_κ where κ is strongly inaccessible and Gödel's model L. For the first we must assume ZFC and the existence of a strongly inaccessible cardinal; for the second, since we are constructing a class and not a set model, only ZFC need be assumed.

SECTION 6.1. A SET MODEL FOR ZFC

An infinite cardinal κ is said to be a strong limit iff, for all $\lambda < \kappa$, $2^\lambda < \kappa$. An uncountable regular strong limit is called a strongly inaccessible cardinal. Notice that a strong limit cardinal is in fact a limit: If $\kappa = \lambda^+$ for some λ, then $2^\lambda \geq \kappa$. Hence a strongly inaccessible cardinal is also weakly inaccessible.

Strongly inaccessible cardinals are often simply called inaccessible.

Theorem 1. Let κ be strongly inaccessible. Then V_κ is a model of ZFC.

The proof of theorem 1 proceeds by a series of lemmas.

Lemma 2. For all ordinals α, V_α is a model of extensionality, regularity, and union.

Proof. Each V_α is transitive and every transitive set models extensionality; every set models regularity; and if $x \in V_\alpha$, so is $\bigcup x$.

Lemma 3. Let α be a limit ordinal. Then V_α models pairing and power set.

Proof. If α is a limit, and x, y are elements of V_α, then there is some $\beta < \alpha$ with $x, y \in V_\beta$. Hence $\{x, y\} \in V_{\beta+1}$, and $\mathcal{P}(x) \in V_{\beta+2}$. Since $V_{\beta+1} \subset V_{\beta+2} \subset V_\alpha$, V_α is a model of pairing and power set.

Note that we have in fact proved something stronger: If α is a limit, then V_α is closed under pairing and power set (i.e., if x, y are elements of V_α, then so are $\{x, y\}$ and $\mathcal{P}(x)$).

Lemma 4. Let α be a limit ordinal. Then V_α models choice.

Proof. If α is a limit, then V_α satisfies the hypothesis of theorem 42 in chapter 3 and hence is a model of choice.

Lemma 5. Every transitive model closed under power set models separation. Hence if α is a limit, V_α models separation.

Proof. Suppose X is transitive and X is closed under power set. Then if $y \in X$ and Φ is a formula, $y_\Phi = \{a \in y: \Phi(a)\} \in \mathcal{P}(y) \in X$. Hence, since X is transitive, $y_\Phi \in X$.

Lemma 6. If $\alpha > \omega$, then V_α models infinity.

Proof. By induction, each $n \in V_{n+1}$, hence $\omega \subset V_\omega$, hence $\omega \in V_\alpha$ for each $\alpha > \omega$.

Lemma 7. If κ is a regular strong limit cardinal, then each element of V_κ has cardinality less than κ, and $|V_\kappa| = \kappa$.

Proof. For the first part it suffices to show that if $\alpha < \kappa$ then $|V_\alpha| < \kappa$. So suppose $\alpha < \kappa$ and suppose we know that for all $\beta < \alpha$, $|V_\beta| < \kappa$. If $\alpha = \beta + 1$ for some β, then $|V_\alpha| = 2^{|V_\beta|} < \kappa$, by induction hypothesis and since κ is a strong limit. If α is a limit, then $V_\alpha = \bigcup V_\beta$, hence $|V_\alpha|$ is the limit of fewer than κ ordinals of size less than κ. So by regularity of κ, $|V_\alpha|$ must have size less than κ.

For the second part: V_κ is an increasing union of κ many sets, so it has size at least κ, but it is the increasing union of sets of size less than κ, so its size is at most κ.

Lemma 8. If κ is a regular strong limit cardinal, then V_κ models replacement.

Proof. Suppose κ is a regular strong limit cardinal. Let Φ be as in the hypothesis of the axiom of replacement: If $\Phi(x, y)$ and $\Phi(x, z)$, then $y = z$, field $(\Phi) \subset V_\kappa$, and all parameters of Φ are elements of V_κ. Let $A \in V_\kappa$. Since κ is a regular strong limit, $|A| < \kappa$. Hence the set $B = \{y: \exists x \in A \, (y = \Phi(x))\}$ has size less than κ. Since field $(\Phi) \subset V_\kappa$, $B \subset V_\kappa$. We show that $B \in V_\kappa$.

Let $f: B \to \kappa$ be defined by $f(y) = \text{rank } y$. Each $f(y) < \kappa$, and there are fewer than κ of them, so range f has a supremum $\alpha < \kappa$. Hence $B \subset V_\alpha$, so $B \in V_{\alpha+1}$, so $B \in V_\kappa$.

Theorem 9

(a) V_ω is a model of all axioms of ZFC except infinity.

(b) If κ is strongly inaccessible, then V_κ models ZFC.

Proof. For (a): ω is a regular strong limit. For (b): A strongly inaccessible cardinal is an uncountable regular strong limit. In particular it is bigger than ω.

Notice that by the completeness theorem (see section 1.6), theorem 9 shows that if $Z^* = $ ZFC minus the axiom of infinity then ZFC proves the consistency of Z^*.

Corollary 10. The following is not a theorem of ZFC: There is a strongly inaccessible cardinal.

We give two proofs. The first is a simple use of the second incompleteness theorem. The second, while more complicated, serves as an introduction to two notions: a more sophisticated use of the "least counterexample" idea than we have seen so far and relativization of concepts from one model to another.

Proof 1. Suppose κ is strongly inaccessible. Then V_κ is a model of ZFC. So "there exists a strongly inaccessible cardinal" implies "ZFC is consistent." But, by Gödel's second incompleteness theorem, ZFC cannot prove its own consistency.

Proof 2. Suppose ZFC does prove the existence of a strongly inaccessible cardinal, and let κ be the least strongly inaccessible cardinal. By theorem 9, V_κ is a model of ZFC, so in V_κ there is an ordinal λ which is, in V_κ, a strongly inaccessible cardinal, i.e.,

(i) If $\alpha < \lambda$, then there is no function in V_κ with domain α whose range is cofinal in λ. (This is the statement that λ is regular in V_κ; note that regularity implies that λ is a cardinal.)

(ii) If $\alpha < \lambda$, then there is no function in V_κ from $\mathscr{P}(\alpha) \cap V_\kappa$ onto λ. (This is the statement that λ is a strong limit in V_κ.)

We now have to show that λ is, in fact, a strongly inaccessible cardinal in V, which will contradict our definition of κ.

If there were, in V, a function from some $\alpha < \lambda$ onto a set cofinal in λ, then, by definition of the V_β's, this function would already exist in V_κ. So λ is regular in V. Similarly, since $\mathscr{P}(\alpha) \cap V_\kappa = \mathscr{P}(\alpha)$ for $\alpha < \kappa$ and since, for $\alpha < \kappa$, any function from $\mathscr{P}(\alpha)$ onto λ would already exist in V_κ, λ is a strong limit in V. Thus λ is strongly inaccessible in V, and we are done.

The next section will indicate that we cannot even prove in ZFC the existence of weakly inaccessible cardinals. A cardinal whose existence cannot be proved in ZFC but whose existence has not been shown to be inconsistent with ZFC is called a large cardinal. Thus weakly inaccessible and strongly inaccessible cardinals are large cardinals. There is an exotic zoo of large cardinals: weakly compact cardinals, measurable cardinals, strongly compact cardinals, huge cardinals, not to mention the n-extendibles, Mahlo's and weakly Mahlo's, and so on. Most of the large cardinals are defined in terms of certain sorts of embeddings from the universe into a proper subset of itself and cannot really be understood without mathematical logic. For an out-of-date reference which remains excellent on the basics, the reader is referred to Drake. Two excellent surveys (also necessarily out-of-date) are the ones by Kanamori, Reinhardt, and Solovay and by Kanamori and Magidor. In the next chapter we will say a little bit about the combinatorics of weakly compact cardinals and measurable cardinals.

SECTION 6.2. THE CONSTRUCTIBLE UNIVERSE

In this section we sketch the construction of and state some facts about the constructible universe L. L was discovered by Gödel as a way to prove that if ZF is consistent then so is ZFC. But it is Jensen's painstakingly detailed techniques that have uncovered so much of the structure not only of L but of classes constructed in similar fashion. The techniques derived from Jensen's work are grounded in concern for the level of complexity of definition of an object and hence cannot be studied without a prerequisite of an advanced mathematical logic course. Here we content ourselves with giving the bare definition of L and stating without proof some facts about it.

Definition 11. Let X be a set and Φ a formula in the language of set theory. We define the formula Φ^X to be the formula derived from Φ by replacing each "$\forall x_i$" occurring in Φ by "$\forall x_i \in X$" and by replacing each "$\exists x_i$" occurring in Φ by "$\exists x_i \in X$."

Definition 12. Let X be a set. We say a set A is definable from X iff there is a formula Φ with parameters in X so that $A = \{x \in X : \Phi^X(x)\}$. The set of all sets definable from X is called $\mathrm{Def}(X)$.

Note: The same formula may define different sets, depending on X. For example, each X is definable from itself via the formula "$x = x$."

Example 13
   (a) The set of even numbers is definable from ω.
   (b) For every ordinal α, if $\alpha \subset X$ and X is transitive, then α is definable from X.
   (c) If $a \in X$ and X is transitive, then a is definable from X.
   (d) If $a \in X$, then $\{a\}$ is definable from X.
   (e) If $a, b \in X$, then $\{a, b\}$ is definable from X.
   (f) If $a \in X$ and X is transitive, then $\bigcup a$ is definable from X.

Proof. For (a) consider the formula "$\exists y \ (x = y + y)$." For (b), if $\alpha \in X$, consider the formula "$x \in \alpha$." Otherwise $\alpha = \mathrm{ON} \cap X$, so consider the formula "x is an ordinal." For (c) consider the formula "$x \in a$." For (d) consider the formula "$x = a$." For (e), consider the formula "$x = a$ or $x = b$." For (f), consider the formula "$\exists y \ (x \in y \in a)$."

Definability leaves a lot out.

Theorem 14. For every infinite set x there is some $y \subset x$, y is not definable from x.

Proof. Since a formula in $\mathrm{Def}(x)$ is a finite sequence whose elements are either symbols in the language or parameters in x, the number of formulas which can define sets in $\mathrm{Def}(x)$ is $|x|$, hence $|\mathrm{Def}(x)| \le |x| < 2^{|x|}$. So some element of $\mathcal{P}(x)$ is not an element of $\mathrm{Def}(x)$; in fact most of them are not elements of $\mathrm{Def}(x)$.

Let us define L.

Definition 15. $L_0 = \emptyset$. If α is a limit ordinal, then $L_\alpha = \bigcup_{\beta < \alpha} L_\beta$. If $\alpha = \beta + 1$, then $L_\alpha = \mathrm{Def}(L_\beta)$. $L = \bigcup_{\alpha \in \mathrm{ON}} L_\alpha$.

Notice the lack of ambiguity here, compared to the definition of the V_α's. Once you know x, $\text{Def}(x)$ will not vary from model to model, but $\mathcal{P}(x)$ might.

We leave it to the reader to show by induction that $L_\alpha \in L_\beta$ iff $\alpha < \beta$, and give as a sample

Proposition 16. Each L_α is transitive.

Proof. Suppose L_β is transitive for each $\beta < \alpha$. If α is a limit, we are done, so suppose $L_\alpha = \text{Def}(L_\beta)$ for some β. Then if $y \in x \in L_\alpha$, $y \in L_\beta$, hence by Example 13(c), $y \in \text{Def}(L_\beta) = L_\alpha$.

Proposition 17. Every ordinal is an element of L.

Proof. It suffices to show that every ordinal is a subset of L. By induction, suppose $\beta \subset L$ for every $\beta < \alpha$. If α is a limit, we are done. If $\alpha = \beta + 1$ for some β, then $\beta \in L_\gamma$ for some γ, hence $\{\beta\} \in L_{\gamma+1}$, so $\alpha \subset L_{\gamma+1}$.

Gödel showed that L is a model of ZFC and GCH. We content ourselves with proving some simple fragments of this theorem.

Proposition 18. L is a model of pairing, union, and power set.

Proof. For pairing and union, by example 13 if $a, b \in L_\alpha$, then $\{a, b\} \in L_{\alpha+1}$ and $\bigcup a \in L_{\alpha+1}$. For power set, if $a \in L$, then by the axiom of replacement there is some ordinal β so that $\mathcal{P}(a) \cap L \subset L_\beta$. (Note the use of class notation here.) The formula "$x \subset a$" then makes $\mathcal{P}(a) \cap L$ an element of $L_{\beta+1}$.

An important consequence of GCH is

Proposition 19. Assume GCH. Then every weakly inaccessible cardinal is strongly inaccessible.

Proof. Suppose κ is a limit cardinal, and $\lambda < \kappa$. By GCH, $2^\lambda = \lambda^+ < \kappa$. Hence κ is a strong limit. So if κ is regular, it is strongly inaccessible.

So a corollary of the fact that L is a model of GCH is

Corollary 20. A weakly inaccessible cardinal in L strongly inaccessible in L.

For the proof of the next proposition we need a bit of notation. If λ is a cardinal in L, then by $(\lambda^+)^L$ we denote the smallest ordinal α in L for which there is no function $f \in L$ with domain $f = \lambda$ and range $f = \alpha$. That is, $(\lambda^+)^L$ is the ordinal that L thinks is λ^+. Notice that $(\lambda^+)^L \leq \lambda^+$ if λ is a cardinal in V. Otherwise, $(\lambda^+)^L \leq |\lambda|^+$.

Proposition 21. If κ is weakly inaccessible, then it is strongly inaccessible in L.

Proof. Since $L \subset V$, if there is no function in the universe from a smaller ordinal onto a set cofinal in κ, there is certainly no such function in L. So a regular cardinal in V is regular in L. Similarly, since any $(\lambda^+)^L \leq \lambda^+$, a limit cardinal in V is a limit cardinal in L. So a weakly inaccessible cardinal in V is weakly, hence strongly, inaccessible in L.

An immediate corollary of proposition 21 and the proof that L models ZFC and GCH is

Proposition 22. If κ is weakly inaccessible, then L_κ is a model of ZFC and GCH.

By propositions 21 and 22, weak inaccessibility is a large cardinal property.

Gödel also proved

Proposition 23. Every class containing all ordinals which is a model of ZF contains L as a subclass.

The main use of L is that it models not only GCH but many other useful combinatorial statements, where useful means "can be used to show a wide variety of mathematical statements consistent." The question becomes "docs $V = L$?"

The statement "$V = L$" is known as the axiom of constructibility. It is called an axiom not because there is general agreement it is true—I doubt that anyone would make a claim for its philosophically self-evident nature. Instead, the word axiom is used to indicate a statement which makes a fundamental claim about the nature of the mathematical universe, a claim which cannot be refuted within ZFC, whether or not we believe that claim to be true. CH is such an axiom. Another such axiom we will meet up with in the next chapter is Martin's axiom.

Since all the axioms of ZFC hold in L, and "$V = L$" holds in L, the axiom of constructibility cannot be refuted in ZFC (just work inside the little piece of the universe we call L—how could you know anything else existed?). But are there models in which the axiom of constructibility fails?

Since L is a model of CH, it suffices to find a model in which CH fails, which Cohen did in 1963. So the axiom of constructibility is independent. Cohen's method, which has been explosively developed in the years since his work, is called forcing. It too is based on ideas of mathematical logic and cannot be included in this book. Some basic references are Burgess, Jech (1971, 1978), and Kunen.

The technique of forcing, which necessarily adds sets to L, gives an intuition (which can neither be proved nor disproved) that L is small. Assuming certain large cardinal hypotheses we can prove that L is very small. A useful large cardinal hypothesis is "0# exists." The definition of 0# is highly model-theoretic: 0# is defined to be a particular set of formulas which completely describe a certain kind of model of set theory. The existence of this sort of model turns out to be much stronger than the statement that set theory is consistent. "0# exists" is implied by some large cardinal hypotheses (e.g., "there exists a measurable cardinal") and implies still others (e.g., "there is a weakly inaccessible cardinal," to name the weakest). For example, Silver proved

Theorem 24. If 0# exists, then ω_1 is inaccessible in L.

Note that the ω_1 in theorem 24 is the real ω_1, not the ordinal that L thinks is ω_1. Thus if 0# exists, L is so small that it thinks the puny cardinal ω_1 is a great big strongly inaccessible cardinal. Recall that the ordinal that L thinks is ω_1 is denoted ω_1^L, and is the first ordinal with no 1–1 function $f \in L$, so that $f: \omega_1^L \to \omega$. A corollary to theorem 24, then, is

Corollary 25. If 0# exists, then ω_1^L is countable.

Another expression of the extreme smallness of L under large cardinal hypotheses is Jensen's covering theorem, stated in the form: The failure of "0# exists" is equivalent to L's being reasonably large.

Theorem 26 (Jensen's Covering Theorem). 0# does not exist iff every uncountable set of ordinals x is contained in a set of ordinals $y \in L$ where $|x| = |y|$.

Corollary 27 (Jensen' Singular Cardinals Theorem). If 0# does not exist, then for every singular strong limit cardinal κ, $2^\kappa = \kappa^+$.

Sketch that Corollary 27 Follows from Theorem 26. By exercise 28 in chapter 5, it suffices to show that $\kappa^{\text{cf}(\kappa)} = \kappa^+$. So suppose $x \subset \kappa$, $|x| =$

$cf(\kappa) = \lambda < \kappa$ where κ is a singular strong limit. By the covering theorem, $x \subset y_x$ for some $y_x \in L$ where $|y_x| \le \lambda^+ < \kappa$ (if $\lambda > \omega$, we have $|y_x| = \lambda$), and $y_x \subset \kappa$. Since GCH holds in L there are at most κ^+ many such y_x and each $\mathcal{P}(y_x) < \kappa$ by hypothesis. Recall that $(\kappa^+)^L \le \kappa^+$. So κ has at most κ^+ many subsets of size λ, hence κ^λ has size no greater than κ^+; on the other hand, κ^λ has size at least κ^+ and we are done.

The ideas of theorem 26 have been extended and applied to other classes which resemble L in the canonical way in which they are constructed.

The nonexistence of inner models of measurable cardinals is a sufficient hypothesis in corollary 27.

EXERCISES FOR CHAPTER 6

1. Consider $V_{\omega+1}$. Which axioms of ZFC is it a model of? What about $V_{\omega+\omega}$? V_{ω_1}?

2. Show that there is some subset of ω in L which is not definable from ω.

3. Show that each finite n is definable from a formula with no constants.

4. (a) Show that if $a, b \in x$ and x is transitive, then $a \times b \in \text{Def}(x)$.
   (b) Give an example where $a, b \in x$ but $a \times b \notin \text{Def}(x)$.

5. Show that $L_\alpha \in L_\beta$ iff $\alpha < \beta$.

6. Show that each $L_\alpha \subset V_\alpha$.

7. Show that the axioms of extensionality, separation, and replacement hold in L.

8. Show that every cardinal in V is a cardinal in L.

9. Assume that $0\#$ does not exist. Let $\lambda = (\kappa^+)^L$ and suppose $cf(\lambda) > \omega$. Show that $\lambda = \kappa^+$.

7

INFINITE COMBINATORICS

INTRODUCTION

In this chapter we use the techniques and ideas already developed to explore some questions in infinite combinatorics. In no way are the areas chosen exhaustive of this enormous topic. We have chosen a few areas which are both fairly easy to describe and which have easily described consequences for the rest of mathematics. These areas are partition calculus, trees, CH, Martin's axiom, stationary sets and Jensen's principle $\diamond$, and measurable cardinals. The sections in this chapter are largely but not completely independent of each other. A reader who knows some finite combinatorics and graph theory should find some of the concepts in the first two sections familiar; a reader who knows some measure theory should find the concepts of the third section familiar.

There is a grand theme in this chapter, and that is the interrelatedness of the various topics. Thus, trees and partition calculus lead to large cardinals; trees are used to understand the real line; and the combinatorial principles CH, MA, and $\diamond$ shed different light on the same combinatorial issues.

SECTION 7.1. PARTITION CALCULUS

Here is a party trick: If you have at least six people together in a room either three of them knew each other previously, or three of them were strangers to each other previously. Why is this? Suppose no three of them knew each other previously. Pick out a person, call him Murgatroyd, and the rest of the group divides into two sets: the people Murgatroyd knew

111

previously, call it class I, and the people Murgatroyd had never seen before, call it class II. If two people in class II had never seen each other before, say Jane and Sally, then we are done, since Murgatroyd, Jane and Sally had all been strangers to each other previously. So we may assume that all the people in class II knew each other before, hence it must have size at most two. So there are at least three people in class I. Murgatroyd knows all the people in class I, so if two of them knew each other we would have three people who all knew each other, a contradiction. Otherwise no two people in class I knew each other, and we are done.

This party trick is the simplest example of a class of theorems known as Ramsey theorems, after the brilliant English mathematician who discovered them and died tragically young. To generalize this party trick it helps to develop some notation; this notation in turn inspires variations which give rise to some of the most difficult, interesting, and useful, as well as some of the most arcane, concepts in infinite combinatorics. Since all of these ideas have to do with partitions, their study is known as partition calculus.

We will develop the notation by analyzing the party trick.

We have a property of pairs of people: Either they know each other or they do not. Thus we really have a partition of $[X]^2 = \{\{x, y\}: x, y \in X$ and $x \neq y\}$. (In general, given a set X we define $[X]^\kappa$ to be the set of subsets of X each of which has size κ; $[X]^{<\kappa}$ is the set of subsets of X each of which has size strictly less than κ.) This partition of $[X]^2$ has two pieces: "know each other," and "do not know each other." We are asking: Can you find a subset of three elements all of whose pairs lie in the same piece of the partition? Such a subset is known as a homogeneous set.

Definition 1

(a) Recall from chapter 1: $\{P_i: i \in I\}$ is a partition of a set Z iff $\bigcup_{i \in I} P_i = Z$ and the P_i's are pairwise disjoint. Note that some of the P_i's can be empty.

(b) If $[X]^p$ is partitioned into sets $\{P_i: i \in I\}$, then $Y \subset X$ is homogeneous for the partition iff, for some i, $[Y]^p \subset P_i$.

Thus the party trick can be rephrased as follows: If $|X| \geq 6$, and $[X]^2$ is partitioned into two sets, then there is some Y homogeneous for the partition, $Y \subset X$, $|Y| \geq 3$.

More compactly, we write this as $6 \rightarrow (3)^2_2$, meaning that if you partition the pairs (this is the upper 2) of a set of six elements into two pieces (this is the lower 2) you will have a homogeneous subset with at least three elements.

Let us give some examples of homogeneous sets.

Example 2

(a) If we partition $[\omega]^2$ into two pieces, the pairs whose product is even, and the pairs whose product is odd, then all sets consisting solely of even numbers are homogeneous. A set which contains at least two odd and at least one even element cannot be homogeneous.

(b) Partition the three element subsets of $\mathbb{R}^2$ according to whether the elements are collinear or not. Then Y is a homogeneous subset iff Y is a subset of some straight line or no three elements of Y are collinear. Thus straight lines are homogeneous, circles are homogeneous, parabolas are homogeneous, but cubics are not.

Returning to our task of generalizing the arrow notation, we write $\kappa \rightarrow (\lambda)^\rho_\sigma$ iff for every partition of $[\kappa]^\rho$ into σ pieces, there is a homogeneous subset of size λ, i.e., a set $A \in [\kappa]^\lambda$ where $[A]^\rho$ is contained in a single element of the partition. Ramsey's theorem for finite sets is that for all finite j, m, k there is an n with $n \rightarrow (j)^m_k$.

The infinite version of Ramsey's theorem is

Theorem 3. For all finite n and m, $\omega \rightarrow (\omega)^n_m$.

Before giving the proof of Ramsey's theorem, we will give an application of it and a few easy facts involving arrow notation.

Theorem 4. Every infinite partial order has either an infinite antichain or an infinite set of pairwise compatible elements.

Proof. Let A be an infinite countable subset of the given partial order, and let $[A]^2 = P_1 \cup P_2$ where $\{x, y\} \in P_1$ iff x and y are incompatible; otherwise $\{x, y\} \in P_2$. Let H be an infinite homogeneous set for this partition. If $[H]^2 \subset P_1$, then H is an infinite antichain; if $[H]^2 \subset P_2$, then H is an infinite set of pairwise compatible elements.

The exercises give several more applications of Ramsey's theorem.

Proposition 5. For all cardinals $\kappa, \lambda, \rho, \sigma, \tau$, if $\kappa \rightarrow (\lambda)^\rho_\sigma$, then

(a) If $\tau > \kappa$, then $\tau \rightarrow (\lambda)^\rho_\sigma$.

(b) If $\tau < \lambda$, then $\kappa \to (\tau)^\rho_\sigma$.

(c) If $\tau < \sigma$, then $\kappa \to (\lambda)^\rho_\tau$.

Proof. For (a): Given a partition $\{P_\alpha : \alpha < \sigma\}$ of $[\tau]^\rho$ we consider the auxiliary partition $\{P^*_\alpha : \alpha < \sigma\}$ where $P^*_\alpha = P_\alpha \cap [\kappa]^\rho$. The homogeneous set for the P^*_α's will work for the P_α's as well.

For (b): Given a homogeneous set of size λ, cut it down to one of size τ.

For (c): Any partition into τ sets is also a partition into σ sets, where the extra sets are simply declared to be empty.

Note that for each infinite κ, $\kappa \to (\kappa)^1_2$ and $\kappa \to (\kappa)^2_1$. Thus the simplest nontrivial arrow relation on an infinite κ is $\kappa \to (\kappa)^2_2$. Let us prove some results about this relation.

Proposition 6. If κ is infinite, and $\kappa \to (\kappa)^2_2$, then $\kappa \to (\kappa)^2_m$ for all $m \in \omega$.

Proof. We work by induction on m. If $\kappa \to (\kappa)^2_{m-1}$, then, given a partition $P_1, \ldots, P_m$ of $[\kappa]^2$ into m pieces let $P'_1 = P_1 \cup P_2$ and for $i > 1$ let $P'_i = P_{i+1}$. The P'_i's form a partition of size $m-1$, so κ has a homogeneous set Y of size κ. If $[Y]^2 \subset P'_i$ for $i > 1$, we are done. Otherwise $[Y]^2 \subset P_1 \cup P_2$ and by invoking $\kappa \to (\kappa)^2_2$ we can find a homogeneous subset of Y of size κ.

It is in fact true that if $\kappa \to (\kappa)^2_2$ then $\kappa \to (\kappa)^n_\alpha$ for every finite n and every ordinal $\alpha < \kappa$, but we will not prove it. Instead we prove

Theorem 7. If $\omega \to (\omega)^2_2$, then for all finite n, m, $\omega \to (\omega)^n_m$.

Ramsey's theorem will then be complete by proving

Theorem 8. $\omega \to (\omega)^2_2$.

In fact, the proof of theorem 8 is an easier version of the proof of theorem 7, so we do it first.

Proof of Theorem 8. Suppose we have partitioned $[\omega]^2$ into two pieces, P_0 and P_1. We recursively build a sequence of natural numbers $\{k_j : j < \omega\}$ and a sequence of infinite sets $\{A_j : j < \omega\}$ so that each $A_{j+1} \subset A_j$, each $k_j \in A_j$, and for each j there is some i with $A_{j+1} \subset \{m : \{k_j, m\} \in P_i\}$. How do we do this? Let $k_0 = 0$ and $A_0 = \omega$. If we have A_j, k_j, $j \leq m$, satisfying the above requirements, notice that there is some

i so that $A = \{k: k > k_m, \ k \in A_m \text{ and } \{k_m, k\} \in P_i\}$ is infinite. Let $A_{m+1} = A$, $k_{m+1} \in A_{m+1}$, and continue.

$$
\begin{array}{ccc}
\cdot & \cdot \cdot \cdot & \cdot \cdot \ \cdot \cdot \cdot \\
\underline{} & \underline{} \\
k_0 & A_1
\end{array}
$$

$$
\begin{array}{ccc}
\cdot \quad \cdot & \quad \cdot \cdot \cdot & \cdot \cdot \\
 & \underline{} & \underline{} \\
k_0 \quad k_1 & A_2
\end{array}
$$

Now that we have the k_n's, define $f(n) = i$ iff for all $m \in A_{n+1}$, $\{k_n, m\} \in P_i$. The function f is defined for all n, and since its range is $\{0, 1\}$ it is constantly equal to some i on some infinite set B. But then if n, m are elements of B, $\{k_n, k_m\} \in P_i$, so $\{k_n: n \in B\}$ is homogeneous. Theorem 8 is proved.

Proof of Theorem 7. By proposition 6 it suffices to fix m and work by induction on n. Suppose $\omega \rightarrow (\omega)^n_m$ and suppose $\mathscr{P} = \{P_1, \ldots, P_m\}$ is a partition of $[\omega]^{n+1}$. For each $k \in \omega$ let $P^k_i = \{s \in [\omega]^n: s \cup \{k\} \in P_i\}$ and let $\mathscr{P}^k = \{P^k_1, \ldots, P^k_m\}$. Note that each $\mathscr{P}^k$ is a partition of $[\omega - \{k\}]^n$. We now construct k_j, A_j as in the proof of theorem 8, with the additional requirement that each A_j be homogeneous for each $\mathscr{P}^{k_r}$, $r \le j$. How can we do this? Given A_j, $k_j \in A_j$, we let $B^j_0 \subset A_j$ be homogeneous for $\mathscr{P}^{k_0}$, $B^j_1 \subset B^j_0$ be homogeneous for $\mathscr{P}^{k_1}$, $B^j_2 \subset B^j_1$ be homogeneous for $\mathscr{P}^{k_2}$, and so on. Then B^j_j is homogeneous for all $\mathscr{P}^{k_r}$, $r \le j$, and we let $A_{j+1} = B^j_j$.

Now we assign $f(l) = i$ iff $A_{l+1} \subset P^{k_l}_i$, note that there is some i so that $B = \{l: f(l) = i\}$ is infinite, and again note that $\{k_l: l \in B\}$ is homogeneous for the original partition $\mathscr{P}$.

Our goal now is to show that, in fact, $\kappa \rightarrow (\kappa)^2_2$ is a large cardinal property, i.e., $\kappa \rightarrow (\kappa)^2_2$ implies that κ is strongly inaccessible for uncountable κ. First we will show that κ is regular; then we will show that κ is a strong limit. On the way to the second goal we will prove a negative partition relation and a fact about the lexicographic order of independent interest.

Theorem 9. If $\kappa \rightarrow (\kappa)^2_2$, then κ is regular.

Proof. Let κ_α, $\alpha < \mathrm{cf}(\kappa)$, be an increasing sequence of cardinals cofinal in κ. For each ordinal $\beta < \kappa$ we write $f(\beta) = $ least α with $\beta < \kappa_\alpha$ and write $\beta \sim \gamma$ iff $f(\beta) = f(\gamma)$. Then $\sim$ is an equivalence relation dividing $[\kappa]^2$ into two pieces: P_1 consists of all those pairs having relation $\sim$, P_2 consists of all those pairs that do not. Since each $\kappa_\alpha < \kappa$, no homogeneous set Y of size κ can have $[Y]^2 \subset P_1$. Hence a homogeneous set of size κ picks at most one element out of every equivalence class for

$\sim$, hence at most one element between each κ_α and $\kappa_{\alpha+1}$. Hence $cf(\kappa) = \kappa$.

Theorem 10. The lexicographic order on the set $^\lambda 2$ has no increasing chains ordered in type λ^+ and no decreasing chains inversely ordered in type λ^+.

Proof. Suppose we had $F = \{f_\alpha : \alpha < \lambda^+\}$, each $f_\alpha \in {}^\lambda 2$, and if $\alpha < \beta$, then $f_\alpha <_L f_\beta$, where $\leq_L$ is the lexicographic order. For each $f \in F$ define $d(f)$ to be the least $\gamma < \lambda$ so that, for some $g \in F$, $f(\gamma) < g(\gamma)$ and $f \mid \gamma = g \mid \gamma$. Note that $f(d(f)) = 0$ for all f. Let $F_\gamma = \{f \in F : d(f) = \gamma\}$. If $f \in F_\gamma$, $g \in F$, let us say that g is a witness for f if $f \mid \gamma = g \mid \gamma$, and $f(\gamma) < g(\gamma)$. Notice that such $g \notin F_\gamma$.

Subclaim. Suppose $f \in F_\gamma$, g is a witness for f. Then $h <_L g$ for every $h \in F_\gamma$.

Proof of Subclaim. Suppose not. Then there is $h \in F_\gamma$ with $h >_L g$, hence $h >_L f$. Since $f \in F_\gamma$, $f \mid \gamma = h \mid \gamma$. What is $h(\gamma)$? If $h(\gamma) = 0$, $h <_L g$, a contradiction. So $h(\gamma) = 1$. But then $h \notin F_\gamma$, a contradiction.

Returning to the proof of theorem 10, each $F_\gamma \subset \{f_\alpha : \alpha < \delta\}$ for some $\delta < \lambda^+$, where f_δ is a witness for some $f \in F_\gamma$. Thus each $|F_\gamma| \leq \lambda$, and, since $F = \bigcup_{\gamma < \lambda} F_\gamma$, $|F| = \lambda$, a contradiction.

A similar proof shows that there is no $\{f_\alpha : \alpha < \lambda^+\}$, each $f_\alpha \in {}^\lambda 2$, and if $\alpha < \beta$ then $f_\alpha >_L f_\beta$.

Theorem 11. For all infinite cardinals, λ, $2^\lambda \not\to (\lambda^+)_2^2$.

Proof. Let $\{f_\alpha : \alpha < 2^\lambda\}$, list the elements of $^\lambda 2$, and again let $\leq_L$ be the lexicographic order. We partition $[^\lambda 2]^2$ into two pieces: $P_1 = \{\{\alpha, \beta\} : \alpha < \beta$ and $f_\alpha <_L f_\beta\}$ and $P_2 = \{\{\alpha, \beta\} : \alpha < \beta$ and $f_\alpha >_L f_\beta\}$. A homogeneous set for P_1 is a well-ordered increasing chain; a homogeneous set for P_2 is a decreasing chain inversely well-ordered. By theorem 10, neither we can have size λ^+, and theorem 11 is proved.

Theorem 11 is especially interesting in light of the Erdös–Rado theorem, which we will not prove here.

Theorem 12 (Erdös–Rado). $(2^\lambda)^+ \to (\lambda^+)_2^2$.

Thus we know exactly which cardinals μ satisfy $\mu \to (\lambda^+)_2^2$.
Finally, we show, as promised

Theorem 13. If $\kappa \to (\kappa)^2_2$, then κ is a strong limit.

Proof. Suppose, by contradiction, there is some $\lambda < \kappa$ with $2^\lambda \geq \kappa$. Then $\kappa \to (\kappa)^2_2$ implies $\kappa \to (\lambda^+)^2_2$ which implies $2^\lambda \to (\lambda^+)^2_2$, contradicting theorem 11.

Since theorem 13 says that $\kappa \to (\kappa)^2_2$ is a large cardinal property for uncountable κ, we give it a name.

Definition 14. An uncountable cardinal κ is said to be weakly compact iff $\kappa \to (\kappa)^2_2$.

In the next section, we will see that weakly compact cardinals have other interesting combinatorial properties.

SECTION 7.2. TREES

A tree is a partially ordered set in which the predecessors of each element are well-ordered; i.e., if $\hat{t} = \{s: s < t\}$, then each $\hat{t}$ is well-ordered.

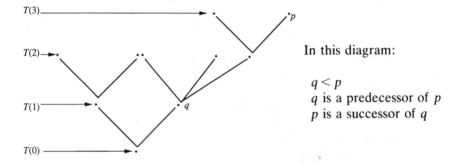

In this diagram:

$q < p$
q is a predecessor of p
p is a successor of q

If T is a tree and $t \in T$ we say the height of t is the order type of $\hat{t}$, i.e., the ordinal isomorphic to $\hat{t}$, and the height of T is the sup of all heights of elements of T. We write $T(\alpha)$ for the set of elements of T of height α and T_α for the set of elements of T of height strictly less than α. $T(\alpha)$ is called the αth level of T; T_α is called an initial segment of T.

For example, if X is a set and κ a cardinal then $T = \bigcup \{{}^\beta X; \beta < \kappa\}$ is a tree of height κ under the ordering $f \leq g$ iff $g \,|\, \mathrm{dom}\, f = f$; each $T(\alpha) = {}^\alpha X$; and each $T_\alpha = \bigcup \{{}^\beta X: \beta < \alpha\}$. If $X = 2$, we call this tree the binary tree of height κ; if $X = \kappa$, we call this the κ-branching tree of height κ.

Note that every subset of a tree is a tree under the induced order.

A branch of a tree is a maximal linearly ordered subset. For example, an element of $^\kappa 2$ is the union of a branch of the binary tree of height κ. An antichain in a tree is a collection of elements which are pairwise incomparable. For example, a level of a tree is an antichain. Another example: In the binary tree of height ω if σ_n is the function with domain $n + 1$ whose value below n is 0 and whose value at n is 1 then $\{\sigma_n : n \in \omega\}$ is an antichain.

In an abuse of notation, we say that two elements of a tree are incompatible iff they are incompatible under $\leq^{-1}$, where $p \leq^{-1} q$ iff $q \leq p$.

The main subject we will consider in this section is the relation between the length of a tree's branches and the size of its levels or of its antichains. These rather innocent-looking questions will quickly involve us in consistency results and large cardinals. First, a basic lemma.

Lemma 15. Two elements in a tree are incomparable under $\leq$ iff they are incompatible (recall that incompatible is defined as incompatible under $\leq^{-1}$).

Proof. Note that two elements are comparable under $\leq$ iff they are comparable under $\leq^{-1}$. Incompatibility always implies incomparability in a partial order, so we need to prove only one direction. Suppose t, s are not incompatible. Then there is some r with $s, t \in \hat{r}$. But $\hat{r}$ is linearly ordered, so s, t are comparable.

That compatibility under $\leq^{-1}$ = comparability is a major aspect of trees; in fact, it serves to define them in the finite case.

Our main combinatorial concern will be

Definition 16. A cardinal κ has the tree property iff for every tree T whose levels and branches each have size less than κ, T has size less than κ. Equivalently, κ has the tree property iff every tree of size κ whose levels have size smaller than κ has a branch of size κ.

For example, a singular cardinal does not have the tree property: Let κ be singular of cofinality λ, and let $\{\kappa_\alpha : \alpha < \lambda\}$ be an increasing sequence whose union is κ. We let f_α be the function with domain κ_α whose value at each $\beta < \kappa_\alpha$ is constantly equal to α, and let T be the tree of initial segments of the f_α's ordered by inclusion; i.e., $g \in T$ iff there are some β, α with $g = f_\alpha \mid \beta$, and $g \leq h$ iff $h \mid \mathrm{dom}\, g = g$. Then T has height κ, hence size at least κ, each level of T has size $\lambda < \kappa$, but T has no branch of size κ.

We will prove that ω has the tree property, that ω_1 does not have the tree property, and that weak compactness is equivalent to having the tree property and being strongly inaccessible.

Theorem 17 (König). ω has the tree property.

Proof. Suppose T is an infinite tree with every level finite. We must construct an infinite branch. Choose t_0 on level 0 to have infinitely many successors (since T is infinite and $T(0)$ is finite, such a t_0 exists), and let S_0 be the set of successors of t_0, $S_0 = \{s: s > t_0\}$. Note that S_0 is an infinite tree with all levels finite, so we can choose t_1 on the 0th level of S_0; t_1 has infinite many successors in S_0. Let S_1 be the set of successors of t_1. S_1 is an infinite tree with all levels finite, so we pick t_2 in $S_1(0)$; t_2 has infinitely many successors in S_1, let S_2 be the set of successors of t_2 in S_1, and so on. In this way we construct in T an infinite sequence $t_0 < t_1 < t_2 < t_3 \ldots$, and by extending to a maximal linearly ordered subset we have an infinite branch.

Theorem 17 is known as König's lemma. The König of König's lemma is not the König of König's theorem (theorem 43 in chapter 5).

Theorem 17 is quite handy when trees must be constructed ad hoc in th course of a proof. For example, if you have a tree of sets with all levels finite, ordered so that $s < t$ implies $t \in s$, then you know your tree is finite.

The young Aronszajn, taking a quick vacation from his usual preoccupation of functional analysis, proved

Theorem 18 (Aronszajn). ω_1 does not have the tree property.

Before giving the proof of theorem 18, we give

Definition 19. A tree is Aronszajn iff it has height ω_1 but all levels and branches are countable.

Thus the proof of theorem 18 is exactly the proof that Aronszajn trees exist. In fact we will construct an Aronszajn tree with a special property (called, in somewhat pedestrian fashion, a special Aronszajn tree).

Proof of Theorem 18. We will construct our Aronszajn tree T to have, as elements, subsets of $\mathbb{Q}$ which are well-ordered in the usual ordering on $\mathbb{Q}$; the ordering will be end-extension; i.e., $\sigma < \tau$ iff σ is a proper initial segment of τ. Since the union of a branch will be a well-ordered subset of $\mathbb{Q}$, and since $\mathbb{Q}$ is countable, no branch of T will be uncountable. In our construction we will require that if $\sigma \in T$ then $\sup \sigma \in \mathbb{Q}$ and, to ensure that the construction can continue, we will require that

(*) if $\alpha \leq \beta$, $\sigma \in T_\alpha$, and $q \in \mathbb{Q}$, with $q > 0$, then there is some $\tau \in T(\beta)$, $\tau > 0$, and sup $\tau = $ sup $\sigma + q$.

Property (*), restricted to initial segments of our tree, together with the other requirements on elements of T, will be our induction hypothesis.

Given T_α we will construct all of $T(\alpha)$ at once; thus once a level has something in it nothing more is added, so we can control the size of our levels.

So suppose we have T_α, $\alpha < \omega_1$, where T_α is countable, each element of T_α is a well-ordered set of rationals with rational sup, and, for each positive rational q, if $\beta < \gamma < \alpha$, then for each $\sigma \in T(\beta)$ there is $\tau \in T(\gamma)$ with $\tau > \sigma$ and sup $\tau = $ sup $\sigma + q$. We list T_α as $\{\sigma_i : i < \omega\}$ and proceed as follows to construct $T(\alpha)$.

> *Case* 1. α is a successor, say $\alpha = \beta + 1$. Then for each finite i, and positive rational q, we pick some element $\tau_{i,q}$ of $T(\beta)$ which extends or is equal to σ_i so that sup $\tau_{i,q} - $ sup $\sigma_i < q$, and extend $\tau_{i,q}$ to a well-ordered sequence in $\mathbb{Q}$, $\tau^*_{i,q}$, so that sup $\tau^*_{i,q} = $ sup $\sigma_i + q$. (We can always do this by extending $\tau_{i,q}$ by the element sup $\sigma_i + q$.) Now let $T(\alpha) = \{\tau^*_{i,q} : i \in \omega, q \in \mathbb{Q}\}$.
>
> *Case* 2. α is a limit. Then there is an increasing sequence $\{\alpha_n : n \in \omega\}$ whose union is α. For each positive rational q we pick an increasing sequence $\{p_{q,n} : n \in \omega\}$ of positive rationals converging to q, and for each finite i, n and rational q we pick $\tau_{i,n,q}$ comparable to σ_i and $\tau_{i,n-1,q}$, so that $\tau_{i,n,q} \in T(\alpha_n)$, and if $\tau_{i,n,q} > \sigma_i$, then sup $\tau_{i,n,q} = $ sup $\sigma_i + p_{q,n}$. Then, for fixed i, q, $\tau_{i,q} = \bigcup \{\tau_{i,n,q} : n \in \omega\}$ is a well-ordered subset of $\mathbb{Q}$ whose sup is sup $\sigma_i + q$. Let $T(\alpha)$ be the set of all $\tau_{i,q}$'s.

In both cases the levels constructed are countable and remain so. In both cases the construction continues so that the induction hypothesis is satisfied, and thus the construction does not come to a halt on any countable level. Thus, if $T = \bigcup \{T(\alpha) : \alpha \in \omega_1\}$, T is an Aronszajn tree.

Definition 20. An uncountable tree is said to be special iff it is the countable union of antichains.

Note that the tree we constructed in the proof of theorem 18 is, in fact, special, since, for each rational q, $\{\sigma \in T : $ sup $\sigma = q\}$ is an antichain. Is every Aronszajn tree special?

Definition 21. A Suslin tree is an uncountable tree with no uncountable branches and no uncountable antichains.

Thus a Suslin tree is an Aronszajn tree (see the exercises), and no Suslin tree is a special Aronszajn tree (see the exercises). Note that every uncountable subset of a Suslin tree is a Suslin tree. While Aronszajn trees are real (i.e., their existence can be proved in ZFC), with Suslin trees we are deeply in the realm of the undecidable. To see just how deeply, let us introduce some abbreviations and list some consistency results:

Definition 22

(a) SH is the statement: There are no Suslin trees. (SH stands for "Suslin's hypothesis.")
(b) EATS is the statement: Every Aronszajn tree is special.

Recall that CH is the continuum hypothesis: $2^\omega = \omega_1$. Later in this chapter we will meet MA, or Martin's axiom, and $MA + \neg CH$, or Martin's axiom with the negation of the continuum hypothesis. We are ready to survey what is known about SH and EATS.

Theorem 23

(a) SH is consistent.
(b) $\neg$SH is consistent.
(c) Furthermore, if $V = L$, then $\neg$SH holds.
(d) And, under $MA + \neg CH$, SH holds.
(e) Furthermore, under $MA + \neg CH$, EATS holds.

We will prove (c) and (d) later in this chapter. Note that (c) implies (b) and (e) implies (d) implies (a). Theorem 23 is stated as it is to give some historical perspective. In the early 1960s (b) was independently proved by Jech and Tennenbaum; Solovay and Tennenbaum proved (a), which they and Martin turned into (d); and Jensen proved (c). A few years later Baumgartner, Reinhardt, and Malitz proved (e).

What theorem 23 leaves out is whether $SH + \neg EATS$ is possible, and whether SH is consistent with either CH or 2^ω not regular.

Theorem 24. The following are consistent:

(a) $CH + SH$.

  (b) $SH + \neg EATS$.
  (c) $SH + CH + \neg EATS$.
  (d) $SH + 2^{\omega} = \aleph_{\omega_1}$.

Again, although (c) implies both (a) and (b), theorem 24 is given in this form for historical perspective. Jensen proved (a) in the late 1960s; Shelah proved (b) in the mid-1970s; Schlindwein proved (c) in the mid-1980s; and Laver proved (d) in the mid-1980s.

Now none of this could have been imagined when Suslin, back in the 1920s, came up with his hypothesis. What question was motivating him?

Suslin was not thinking of trees at all, but of lines. In particular, he was thinking of the real line, which has the properties that

  (i) No collection of pairwise disjoint intervals is uncountable, and
  (ii) There is a countable dense subset.

Clearly (ii) implies (i). Which of these properties alone characterize subsets of the real line? We will show below

Theorem 25. A linear order satisfies (ii) iff it is order-isomorphic to a subset of the reals.

But does (i) suffice? Suslin's hypothesis, the way Suslin stated it, was that (i) does suffice, i.e., (i) implies (ii). If Suslin's hypothesis fails, there is a linear order satisfying (i) and not (ii). Such an order is known as a Suslin line, and it clearly cannot embed in a suborder of the reals. The connection between lines and trees showing that both versions of SH are equivalent was worked out independently and with much duplication of effort over the next 20 years, and it is to this that we now turn our attention.

In the discussion that follows we generalize the usual notation of intervals on $\mathbb{R}$ to arbitrary linear orders. Thus, in a linear order X, (x, y) means $\{z: x < z < y\}$; the endpoints of (x, y) are x and y; if $z \in (x, y)$, we call (x, y) an interval around z; $x \in \bar{A}$ iff every interval around x contains an element of A; and A is dense in X iff $X = \bar{A}$. ($\bar{A}$ is called the closure of A.)

Let us prove theorem 25. The first step is an important theorem of Cantor.

Definition 26. A linear order is dense iff for every $x < y$ there is z with $x < z < y$.

Theorem 27 (Cantor)

   (a) Every nonempty countable dense linear ordering without endpoints is order-isomorphic to $\mathbb{Q}$.
   (b) Every countable linear ordering is order-isomorphic to a subset of $\mathbb{Q}$.

Proof

(a) Let X be a countable dense linear ordering without endpoints, and enumerate X as $\{x_n: n \in \omega\}$. Let $\mathbb{Q} \in \{q_n: n \in \omega\}$. We construct $f: X \to \mathbb{Q}$, so $x \le y$ iff $f(x) \le f(y)$ as follows: Let $f(x_0) = q_0$. Suppose at stage k we have defined f on $X_k = \{x_n: n < k\}$ and we have also defined f^{-1} on $\mathbb{Q}_k = \{q_n: n < k\}$.

Let $A_k = X_k \cup f^{-1}[\mathbb{Q}_k]$ be the domain of f at stage k. We want to define $f(x_k)$ and $f^{-1}(q_k)$.

If $x_k \in A_k$, we know $f(x_k)$. Otherwise, suppose $a = \sup\{x \in A_k: x < x_k\}$, $b = \inf\{x \in A_k: x > x_k\}$ are both defined. Since $a < b$, $f(a) < f(b)$. Note that $f[A_k] \cap (f(a), f(b)) = \emptyset$. By density of $\mathbb{Q}$, $(f(a), f(b)) \neq \emptyset$. Pick some q with $f(a) < q < f(b)$. Let $f(x_k) = q$.

The case where either a or b is not defined is left to the reader.

If $q_k \in f[A_k \cup \{x_k\}]$, we are done. Otherwise suppose $a = \sup\{p \in f[A_k]: p < q_k\}$, $b = \inf\{p \in f[A_k]: p > q_k\}$ are both defined. This time $A_k \cap (f^{-1}(a), f^{-1}(b)) = \emptyset$, but, by density, there is some $x \neq x_k$, $x \in (f^{-1}(a), f^{-1}(b))$. Pick such an x and let $f(x) = q_k$.

Again, the case where either a or b is not defined is left to the reader.

(b) Let X be a countable linear ordering. By a countable recursive construction, extend X to a countable dense linear ordering with no endpoints, X^*. Since X^* is order-isomorphic to $\mathbb{Q}$, we are done.

The argument in (a) is Cantor's back-and-forth argument, widely used in model theory, the third method of proof invented by Cantor which we have seen.

A corollary of theorem 27 is

Theorem 28. A linear order with a countable dense set is order-isomorphic to a subset of $\mathbb{R}$.

Proof. Let X be a linear order with a countable dense set D. Then every point in X is the supremum of an initial segment of D. Let D be order-isomorphic to $Y \subset \mathbb{Q}$. Then X is order-isomorphic to a subset of $\bar{Y}$, where $\bar{Y}$ is the closure of Y in $\mathbb{R}$.

Since every subset of $\mathbb{R}$ has a countable dense set, theorem 25 is proved.

Having shown that Suslin's hypothesis for lines makes sense, we are ready to turn our attention to the connection between trees and lines. First, some facts about trees.

Definition 29. A tree is a splitting tree iff above every element are two incomparable elements; i.e., for all t there are $r, s > t$ with r, s not comparable.

Proposition 30. If there is a Suslin tree, there is a Suslin tree which is a splitting tree.

Proof. Let T be a Suslin tree. Let $S = \{t \in T;$ if $r, s > t$, then r and s are comparable$\}$. Let A be the set of minimal elements of S. A is an antichain, so it is countable. If $t \in A$, then $\{r \in T : r > t\}$ is countable, since it is a chain. So $T - \bigcup_{t \in A} \{r \in T : r > t\}$ is an uncountable subtree of T, hence Suslin.

Proposition 31. An uncountable splitting tree is Suslin iff it has no uncountable antichains.

Proof. Suppose a splitting tree has an uncountable branch b. For each $t \in b$ there is some $r_t > t$ with $r_t \notin b$. Then $\{r_t : t \in b\}$ is an uncountable antichain, so the tree is not Suslin.

Now we show how to associate lines to trees and trees to lines.

Proposition 32. Every tree extends to a linear order.

Proof. Suppose T is a tree under the order $\leq_T$. Let $\leq_\alpha$ be an arbitrary linear order on $T(\alpha)$. For $x \in T(\alpha)$ and $\beta < \alpha$ we define $x(\beta)$ to be the unique $z \in T(\beta)$ with $z < x$. We define the linear order on T as follows: $x \leq y$ iff

(i) $x \leq_T y$ or
(ii) if α is the least ordinal with $x(\alpha) \neq y(\alpha)$, then $x(\alpha) \leq_\alpha y(\alpha)$.
It is easy to check that this is a linear order.

An order extending $\leq_T$ constructed as in the proof of proposition 32 will be called a canonical linear extension.

Definition 33. Let X be linearly ordered. A tree of intervals on X is a tree whose elements are intervals in X ordered under the following order: $I \leq J$ iff $J \subset I$, where incomparable intervals are disjoint.

Thus, as we move up in a tree of intervals, the intervals grow smaller.

Theorem 34. There is a Suslin line iff there is a Suslin tree.

Proof. Recall that if there is a Suslin tree then there is a splitting Suslin tree. We need an analogous fact about Suslin lines.

Proposition 35. If there is a Suslin line, then there is a dense Suslin line.

Proof. Suppose X is a Suslin line. If $a, b \in X$, $a < b$, and the interval (a, b) is empty, say that a, b are a bad pair. For each bad pair (a, b), insert a copy of $\mathbb{Q}$ between a and b and let X^* be the resulting linear order. Note that X has a countable dense set iff X^* has one, and that an uncountable disjoint family of open sets in X^* gives rise to a disjoint family of open sets in X of the same cardinality. So X^* must be Suslin.

Let us return to the proof of theorem 34. Assume there is a Suslin line. Then there is a dense Suslin line, so let X be one. We construct a Suslin tree of intervals on X by uncountable recursion. The main step is the following

Claim. Suppose X is a dense Suslin line. Then there is a countable splitting tree of intervals on X, and every countable splitting tree T of intervals on X has an extension $S \neq T$ which is also a countable splitting tree of intervals on X.

Proof. Let $A = \{x : x$ is an endpoint of some interval $J \in T\}$. (If $T = \emptyset$, $A = \emptyset$.) Since T is countable, A is countable. Since no countable set is dense in X, there is some interval I so that $I \cap A = \emptyset$. Hence, for any interval $I_1 \subset I$, and any $J \in T$, either $I_1 \cap J = \emptyset$ or $I_1 \subset J$. Since X is dense, there is a nonempty splitting tree S_I where every $J \in S_I$ is a subinterval of I. Let $S = T \cup S_I$.

Using the claim, we build up an increasing sequence of countable splitting trees $\{T^\alpha : \alpha < \omega_1\}$ and let $T^{\omega_1} = \bigcup_{\alpha < \omega} T^\alpha$. Since T^{ω_1} is an uncountable splitting tree with no uncountable antichains, it is Suslin. Thus the existence of Suslin lines implies the existence of Suslin trees.

For the other direction, let us assume we have a splitting tree T which is Suslin under the order $\leq_T$. Let $B = \{b : b$ is a branch of $T\}$, and let $T^* = T \cup B$. We give T^* the following tree order $\leq^+$:

(i) If $t, s \in T$, then $t \leq^+ s$ iff $t \leq_T s$.
(ii) If $t \in T$ and $b \in B$, then $t \leq^+ b$ iff $t \in b$.
(iii) If $t \in T^*$ and $b \in B$, then $b \not\leq^+ t$ if $b \neq t$.

Now let $\leq$ be a canonical linear extension of $\leq^+$. We show that B under the order $\leq$ is a Suslin line.

It has no countable dense set: Since every branch is countable, any countable set is contained in some countable initial segment T^*_α, and if t is an element of T with height greater than α, by splitting there is some $s \in T$ with $t \leq_T s$ and the interval (t, s) under the ordering $\leq$ has at least three elements of B in it (see picture), hence contains a nonempty interval in B. But since $t <_T s$, any b in $(t, s) \cap B$ cannot be in T^*_α, so T^*_α was not dense.

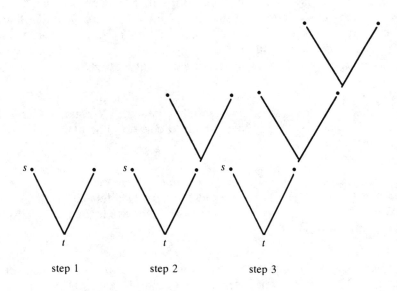

step 1   step 2   step 3

Suppose we have a pairwise disjoint collection C of nonempty intervals in B. Since the endpoints of the intervals in C are in B, the endpoints of each I in C are not comparable in T^*. For each interval $I = (r_I, s_I)$ in C, let $I^* = \{t \in T : r_I < t < s_I\}$. Pick $b(I) \in I$ and $t_I \in b(I) - s_I$. Then $t_I \in I^*$ and if $t_I <_T t_J$, then $t_J \in I^*$. But then $b(J) \in I$, which contradicts the intervals being pairwise disjoint. Hence the t_I's are incomparable in T, hence C is countable.

Theorem 34 was first proved by Kurepa in 1935 and published in a Belgrade journal where it languished unnoticed by the rest of the world. It was rediscovered independently in the 1940s by Sierpinski in Poland and by Miller in the United States.

Now that we have connected our study of trees to questions about the real line, let us connect trees to large cardinals.

Theorem 36. An uncountable cardinal κ is weakly compact iff it is strongly inaccessible and has the tree property.

Proof. Suppose κ is weakly compact. We already know from the last section that it is strongly inaccessible, so it remains to prove that it has the tree property.

Suppose T is a tree of size κ in which each level has size strictly less than κ. List the elements of T as $\{t_\alpha : \alpha < \kappa\}$. Let $\leq_T$ be the tree order on T, and let $\leq$ be a canonical linear extension of it. Let P_0 be the set of all pairs $\{\alpha, \beta\}$ where $\alpha < \beta$ and $t_\alpha < t_\beta$; let P_1 be $[\kappa]^2 - P_0$. Let H be homogeneous for this partition, $|H| = \kappa$, and let $A = \{t : t <_T t_\alpha$ for κ many $\alpha \in H\}$. Since $\{t_\alpha : \alpha \in H\}$ is a subtree of T of size κ and since each level of T has size less than κ, $A \cap T(\alpha) \neq \emptyset$ for each $\alpha < \kappa$. We will show that A is a branch. Suppose not. Then there are two incompatible elements (in the ordering $\leq_T$) t, s in A with $t < s$. For high enough $\alpha < \beta < \gamma$ elements of H, we have $t <_T t_\alpha <_T t_\gamma$ and $s <_T t_\beta$. But then $t_\alpha < t_\gamma$ and $t_\beta \not< t_\gamma$, so H is not homogeneous, a contradiction.

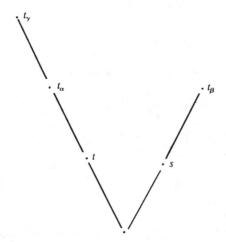

For the other direction, if κ is strongly inaccessible and has the tree property, given a partition of $[\kappa]^2$ into two pieces P_0, P_1, we inductively construct a tree $T = \{t_\alpha : \alpha < \kappa\}$ of functions ordered by end-extensions as follows: $t_0 = \emptyset$. If we have $\{t_\beta : \beta < \alpha\}$, we define t_α inductively as follows: $t_\alpha(0) = i$ iff $\{\alpha, 0\} \in P_i$; for each succeeding γ we ask if, for some $\beta < \alpha$, $t_\alpha \mid \gamma = t_\beta$. If the answer is no, we stop and declare t_α to be defined. If the answer is yes, then β is unique and we define $t_\alpha(\gamma) = i$ iff $\{\alpha, \beta\} \in P_i$. Since elements of T with same height have the same domain, by strong inaccessibility every level of T has size less than κ. So by the tree property, T has a branch B of size κ. Define C_i to be the set of all α where both t_α and $\widehat{t_\alpha i}$, are in B. For some i, C_i has size κ, but if $\alpha < \beta \in C_i$, then $t_\beta \geq \widehat{t_\alpha i}$, so $\{\alpha, \beta\} \in P_i$ and we are done.

A theorem which we will not prove here is that not every strongly inaccessible cardinal is weakly compact. Thus, weak compactness is a stronger large cardinal property than inaccessibility.

SECTION 7.3. MEASURABLE CARDINALS

Measurable cardinals are the last large cardinals we shall talk about; they are also the smallest of the really big cardinals, in a sense we will make precise later. To define them we have to make some further definitions about ultrafilters. Recall from chapter 2 that a proper ultrafilter on a set X is a family $\mathcal{F}$ of subsets closed under supersets and finite intersection, with

(1) $\emptyset \notin \mathcal{F}$.

(2) For every $A \subset X$ either $A \in \mathcal{F}$ or $X - A \in \mathcal{F}$.

We call $\mathcal{F}$ a nonprincipal ultrafilter if, in addition,

(3) Every element of $\mathcal{F}$ is infinite.

Recall that if $\mathcal{F}$ is an ultrafilter, $A \in \mathcal{F}$, and $A = B \cup C$, then either B or C is an element of $\mathcal{F}$.

Definition 37. An ultrafilter $\mathcal{F}$ is said to be κ-closed iff, for every $\mathcal{A} \subset \mathcal{F}$ with $|\mathcal{A}| < \kappa$, $\bigcap \mathcal{A} \in \mathcal{F}$.

Thus, every ultrafilter is ω-closed.

Note that a κ-closed nonprincipal ultrafilter contains no sets of size smaller than κ.

Definition 38. An uncountable cardinal κ is a measurable cardinal iff there is a κ-closed nonprincipal ultrafilter on κ.

Why are these called "measurable?" If you know any measure theory, consider a two-valued measure on $\mathcal{P}(\kappa)$ as follows: Given the κ-closed ultrafilter $\mathcal{F}$ we define the measure μ by $\mu(A) = 1$ if $A \in \mathcal{F}$; $\mu(A) = 0$ otherwise. You can check that this is a measure. It is what is known as a

κ-additive measure, that is, the union of fewer than κ many disjoint sets has measure equal to the sup of the measures of the individual sets. (Note that no two distinct sets in $\mathscr{F}$ are disjoint.)

Theorem 39. Every measurable cardinal is weakly compact.

Proof. Let κ be measurable. We must show that $\kappa \rightarrow (\kappa)_2^2$.

So let $\mathscr{F}$ be a κ-closed ultrafilter on κ, and let $[\kappa]^2 = P_0 \cup P_1$. We imitate the proof of theorem 7 in section 7.1. Let $\alpha_0 = 0$. Since $\mathscr{F}$ is an ultrafilter and $\kappa = \{0\} \cup \{\beta \colon \beta > 0 \text{ and } \{0, \beta\} \in P_0\} \cup \{\beta \colon \beta > 0 \text{ and } \{0, \beta\} \in P_1\}$, there is a unique i with $\{\beta \colon \beta > 0 \text{ and } \{0, \beta\} \in P_i\} \in \mathscr{F}$. Let i_0 be this unique i, and let $A_0 = \{\beta \colon \beta > 0 \text{ and } \{0, \beta\} \in P_{i_0}\}$.

Now suppose, at stage $\eta < \kappa$ we have a descending sequence $\{A_\gamma \colon \gamma < \eta\}$ of elements of $\mathscr{F}$, ordinals α_γ for $\gamma < \eta$, where if $\gamma < \delta < \eta$ then $\alpha_\delta \in A_\gamma$ and $\alpha_\gamma < \inf(A_\gamma)$, and numbers i_γ for $\gamma < \eta$ where $\{\alpha_\gamma, \beta\} \in P_{i_\gamma}$ if $\beta \in A_\gamma$. Consider $\bigcap_{\gamma < \eta} A_\gamma = A$. Since $\mathscr{F}$ is κ-closed, $A \in \mathscr{F}$. We pick $\alpha_\eta \in A$. Since $\alpha_\eta + 1 < \kappa$, and since $\kappa = \sup\{\beta \colon \beta > \alpha_\eta \text{ and } \{\alpha_\eta, \beta\} \in P_0\} \cup \{\beta \colon \beta > \alpha_\eta \text{ and } \{\alpha_\eta, \beta\} \in P_1\} \cup \alpha_{\eta+1}$, there is some $i = i_\eta$ with $A'_\eta = \{\beta \colon \beta > \alpha_\eta \text{ and } \{\alpha_\eta, \beta\} \in P_{i_\eta}\} \in \mathscr{F}$. Let $A_\eta = A'_\eta \cap A$. Finally, note that there are κ many i_γ's that are the same i. Then $\{\alpha_\gamma \colon i_\gamma = i\}$ is the desired homogeneous set.

In fact it is true that if κ is measurable then there are κ many weakly compact cardinals below κ, but we will not prove that.

Measurable cardinals are of particular interest because of the following:

Theorem 40. κ is measurable iff there is a 1–1 map j from the universe V to a transitive subclass M containing all the ordinals where

(1) If Φ is a formula and a_1 through a_n are sets then $\Phi(a_1, \ldots, a_n)$ holds in V iff $\Phi(j(a_1), \ldots, j(a_n))$ holds in M.[†]

(2) j is the identity on V_κ.

(3) $j(\kappa) > \kappa$.

Theorem 40 says that the existence of a measurable cardinal guarantees the existence of a properly smaller copy of the universe. This is what gives measurable cardinals their power. The really big cardinals alluded

---

[†]Let us give an example of $\Phi(j(a_1), \ldots, j(a_n))$. Suppose $\Phi(y)$ is $\forall x(y \in x \rightarrow \exists z(y \in z))$. Then $\Phi(j(y))$ is the formula $\forall x(x \in j(y) \rightarrow \exists z(j(y) \in z))$. The point is that variables which are attached to quantifiers do not have j applied to them.

to in the first paragraph of this section are exactly those which admit more and more powerful sorts of embeddings satisfying properties (1) through (3).

While we do not have the logical machinery in this book to prove theorem 40 in both directions, we can prove one direction and use the map j to show that measurable cardinals do not exist in L.

First, let us explore what property (1) tells us about j.

Proposition 41. Suppose j is a 1–1 map; $j: V \to M$, where M is a transitive subclass of V, M contains all the ordinals, and property (1) holds. Then

(a) For every x, y, $j(x - y) = j(x) - j(y)$.
(b) For every x, $j(\bigcap x) = \bigcap j(x)$.
(c) For every x, y, if $x \subset y$, then $j(x) \subset j(y)$.
(d) If $x = \{y_\alpha : \alpha < \lambda\}$, then $j(x) = \{z_\alpha : \alpha < j(\lambda)\}$, where, if $\alpha < \lambda$, then $z_{j(\alpha)} = j(y_\alpha)$.

Proof. We prove (a), (b), and (d), leaving (c) to the reader.

(a) $x - y$ satisfies $\forall z(z \in x - y$ iff $z \in x$ and $z \notin y)$. So $j(x - y)$ satisfies $\forall z(z \in j(x - y)$ iff $z \in j(x)$ and $z \notin j(y))$. But "$z \in j(x)$ and $z \notin j(y)$" is the definition of $j(x) - j(y)$, so $j(x - y) = j(x) - j(y)$.
(b) The proof is similar to the proof of (a), using the fact that $\bigcap x$ satisfies $\forall y(y \in \bigcap x$ iff $\forall z \in x(z \in y))$.
(d) Now let f be the function $f(\alpha) = y_\alpha$ for $\alpha < \lambda$. Since x is the range of f, $j(x)$ is the range of $j(f)$, and $j(x) = \{z_\alpha : \alpha < j(\lambda)\}$, where $j(f)(\alpha) = z_\alpha$. So if $y = f(\alpha)$, $j(y) = j(f)(j(\alpha)) = z_{j(\alpha)}$.

Theorem 42. Suppose there is a 1–1 map $j: V \to M$, where M is a transitive subclass of V, M contains all the ordinals, and properties (1), (2), and (3) of theorem 40 hold. Let κ be the cardinal of (2) and (3). Then κ is measurable.

Proof. We define $\mathscr{F}$ by $x \in \mathscr{F}$ iff $x \subset \kappa$ and $\kappa \in j(x)$. We need to show that $\mathscr{F}$ is nonprincipal, closed under superset, closed under intersection of fewer than κ elements, and that if $x \subset \kappa$ then either $x \in \mathscr{F}$ or $\kappa - x \in \mathscr{F}$.

$\mathscr{F}$ is nonprincipal: Since $\alpha < \kappa$ implies $\{\alpha\} \in V_\kappa$ implies $j(\{\alpha\}) = \{\alpha\}$, no $\{\alpha\} \in \mathscr{F}$.

$\mathscr{F}$ is closed under superset: If $\kappa \in j(x)$ and $x \subset y$, then, by proposition 41(c), $\kappa \in j(y)$.

If $\lambda < \kappa$ and $\{x_\alpha : \alpha < \lambda\} \subset \mathcal{F}$, then $\bigcap_{\alpha < \lambda} x_\alpha \in \mathcal{F}$: By proposition 41(d), if $X = \{x_\alpha : \alpha < \lambda\}$, then $j(X) = \{j(x_\alpha) : \alpha < \lambda\}$—this is because $j(\lambda) = \lambda$. By proposition 41(b), $j(\bigcap X) = \bigcap j(X)$. So $\bigcap X \in \mathcal{F}$ iff $\kappa \in j(\bigcap X)$ iff $\kappa \in \bigcap j(X)$ iff $\forall \alpha < \lambda (\kappa \in j(x_\alpha))$ iff $X \subset \mathcal{F}$, which is true.

If $x \subset \kappa$, then either $x \in \mathcal{F}$ or $\kappa - x \in \mathcal{F}$: $j(x) \subset j(\kappa)$ and $\kappa \in j(\kappa)$, so either $\kappa \in j(x)$ or $\kappa \in j(\kappa) - j(x) = j(\kappa - x)$ by proposition 41(a).

The next theorem says that measurable cardinals do not exist in L.

Theorem 43 (Scott). If $V = L$, then there are no measurable cardinals.

Proof. Suppose not, and let κ be the smallest measurable cardinal. Let j, V, M be as in theorem 40. Then, in V, κ satisfies the statement "κ is the least measurable cardinal." So, in M, $j(\kappa)$ satisfies the statement "$j(\kappa)$ is the least measurable cardinal." Since $V = L \subset M \subset L$, $M = L$ and κ is measurable in M. Thus $j(\kappa) = \kappa$, a contradiction.

Beyond the scope of this book, but important to mention, is the connection between large cardinals and the Axiom of Determinacy or AD. AD says that certain infinite games have winning strategies, and contradicts AC. There is a rich literature on infinite games and their strategies, which is closely connected to descriptive set theory—descriptive set theory is the study of the classification of sets of reals begun by Borel and continued by the Polish school of Kuratowski and Sierpinski. One of the more remarkable recent developments in set theory has been the connection of descriptive set theory, its roots comingled with the roots of topology and analysis, with large cardinals, beginning with Solovay's theorem that under AD the cardinal ω_1 is measurable, and moving to results spearheaded by Woodin, Steel, Martin, and Shelah, among others, which say that the existence of certain large cardinals implies the existence of an inner model of AD.

SECTION 7.4. CH

In this section we apply CH to some combinatorial questions about $\mathcal{P}(\omega)$, $^\omega\omega$, and ω_1. In the next section we apply Martin's axiom to these same questions.

Recall from chapter 2 the definition of a nonprincipal ultrafilter. Recall from chapter 1 the Frechét order on $^\omega\omega$: We define $f =_* g$ if $\{n : f(n) \neq g(n)\}$ is finite, and define $f \leq_* g$ iff $f = g$ or $\{n : f(n) > g(n)\}$ is finite. While $\leq_*$ is not a partial order on $^\omega\omega$, it defines a partial order on the equivalence classes modulo $=*$ and is called, in an act of generosity, the Frechét order.

The concepts we will work with in this section and the next are defined in

Definition 44

(a) Let $\mathscr{F}$ be an ultrafilter on ω. $\mathscr{A} \subset \mathscr{F}$ is said to be a base for $\mathscr{F}$ iff for every element A of $\mathscr{F}$ there are $A_0, \ldots, A_n \in \mathscr{A}$ with $A \supset \bigcap_{i \leq n} A_i$.

(b) Let $\mathscr{A}$ be a family of subsets of ω. $\mathscr{A}$ has a lower bound iff there is some infinite A with $A - B$ finite for all $B \in \mathscr{A}$.

(c) Let F be a family of functions in $^\omega\omega$. F is said to be dominating iff for every $g \in {}^\omega\omega$ there is some $f \in F$ with $g \leq * f$.

(d) Let F be a family of functions in $^\omega\omega$. F is said to be dominated iff there is some $g \in {}^\omega\omega$ with $g \geq * f$ for each $f \in F$. In this case we say g dominates F.

We will prove

Theorem 45. Assume CH. Then

(a) Every base for a nonprincipal ultrafilter on ω has size 2^ω.

(b) Every dominating family in $^\omega\omega$ has size 2^ω.

The proof of theorem 45 will follow from

Theorem 46

(a) If $\mathscr{A}$ is a countable family of subsets of ω where every intersection of finitely many elements of $\mathscr{A}$ is infinite, then $\mathscr{A}$ has a lower bound.

(b) Every countable family in $^\omega\omega$ is dominated.

Proof That Theorem 45 Follows from Theorem 46. Since CH holds, any subset of $\mathscr{P}(\omega)$ or of $^\omega\omega$ is either countable or has size 2^ω. A base for a nonprincipal ultrafilter cannot have a lower bound A, since then A could be split into two infinite sets neither of which is in the filter; and a dominating family cannot be dominated by a function g, since the function $h(n) = g(n) + 1$ would be a counterexample to the family's being dominating.

Proof of Theorem 46. Theorem 46(a) is immediate from theorem 37 in chapter 1, so we just prove (b).

Let F be a countable collection, $F = \{f_i : i < \omega\} \subset {}^\omega\omega$. We define a dominating g as follows: $g(n) = 1 + \sum_{i<n} f_i(n)$. Then, for each i, $g(n) > f_i(n)$ for all $n > i$, and F is dominated.

In the next section we will prove, under Martin's axiom, analogues of theorem 45. Thus we will have an example of how Martin's axiom and the continuum hypothesis allow us to draw similar conclusions. In contrast, the next theorem fails under $MA + \neg CH$.

Theorem 47 (Erdős–Rado). Assume CH. Then there are H, K with $\omega \times \omega_1 = H \cup K$ so that if $A \in [\omega]^\omega$ and $B \in [\omega_1]^{\omega_1}$ then $A \times B$ is a subset of neither H nor K.

Theorem 47 is an example of the negation of a generalization of a partition relation. This type of generalization has the form: Every time you split an object into a fixed number of pieces, one piece contains a copy of the object. Theorem 47 says that under CH, if the object is $\omega \times \omega_1$, and you split it into two pieces, you need not get a copy of $\omega \times \omega_1$ contained in one piece.

Proof of Theorem 47. Using CH, let $\{a_\alpha : \alpha < \omega_1\}$ enumerate $[\omega]^\omega$. Using a recursive construction of length ω_1, at stage β we will decide which (k, β) go into H and which into K. So at stage β let $\{C_n : n < \omega\}$ enumerate $\{a_\alpha : \alpha < \beta\}$. Since each C_n is infinite, we pick disjoint infinite sets $E, F \subset \omega$ with $C_n \cap E \neq \emptyset$ and $C_n \cap F \neq \emptyset$ for all n. If $k \in E$, put $(k, \beta) \in H$; otherwise put $(k, \beta) \in K$.

Now suppose B is uncountable and A is infinite. Then A is some a_α and there is some $\beta \in B$ with $\beta > \alpha$. But by construction there is some $k \in A$ with $(k, \beta) \in H$, and some $j \in A$ with $(j, \beta) \in K$. So $A \times B \not\subseteq H$ and $A \times B \not\subseteq K$, which is what we wanted to prove.

SECTION 7.5. MARTIN'S AXIOM

Martin's axiom is a statement about partial orders. Before stating it we need some definitions.

Definition 48

(a) A partial order is said to be ccc iff every antichain is countable. (The initials "ccc" stand for "countable chain condition," even though it is antichains that are the subject. The misleading nature of this notation has been noticed and there have been attempts to change it, but tradition outweighs reason, old habits die hard, and "ccc" it remains.)

(b) A set D is dense in a partial order X iff for every $x \in X$ there is some $y \in D$ with $x \geq y$.

Notice that this use of the word "dense" is quite different than the use in section 7.2. For example, $(-\infty, 1)$ satisfies definition 48(b), but it is not dense in $\mathbb{R}$ as a linear order.

Example 49. The partial order of open sets in $\mathbb{R}$ under inclusion is a ccc partial order, since every interval contains a rational, hence every collection of disjoint open sets is countable; the collection of open intervals with rational endpoints is a dense subset of this order.

Example 50. If T is the binary tree on ω and $D = \{f \in T$: for some n dom $f = n + 1$ and $f(n) = 0\}$, then f is dense (given an arbitrary $g \in T$ extend it to a function whose last value is 0); since T is countable, it is ccc.

Another example of a ccc partial order is a Suslin tree under $\leq^{-1}$. This has no countable dense set.

Definition 51. A subset G of a partial order X is said to be a filter if every finite subset of G has a lower bound in G, and if $x \in G$ and $y \geq x$, then $y \in G$.

Notice that a filter on a cardinal κ is exactly a filter on the partial order $\mathcal{P}(\kappa)$, ordered under inclusion.

Notice that every filter in a tree under $\leq^{-1}$ is a chain (because any two elements of a filter are compatible hence, in a tree, comparable under $\leq^{-1}$).

Definition 52. Let $\mathcal{D}$ be a family of dense subsets of some partial order X, and let G be a filter on X. Then G is said to be $\mathcal{D}$-generic iff $G \cap D \neq \emptyset$ for all $D \in \mathcal{D}$. We say that G meets D for all $D \in \mathcal{D}$.

For example, suppose X is a splitting partial order; that is, for every $x \in X$ there are at least two incompatible elements below x. Then for every chain C of X the set $D_C = \{x : x \notin C\}$ is dense. Hence a filter meeting each D_C could not be a chain, and if X is a tree, then there is no filter generic for $\{D_C : C$ a chain$\}$.

The question is: when do generic filters exist? Given a family $\mathcal{D}$ of dense sets, when are we guaranteed the existence of a $\mathcal{D}$-generic filter?

Lemma 53 (Rasiowa–Sikorski). Let X be a partial order and $\mathcal{D}$ a countable family of dense sets. Then there is a $\mathcal{D}$-generic filter on X.

Proof. In fact, the filter will be generated by a chain.

So let $\mathcal{D} = \{D_i : i < \omega\}$. We define $G = \{g_i : i < \omega\}$ as follows: Let $g_0 \in D_0$. Since D_1 is dense there is some $g_1 \in D_1$, $g_1 \leq g_0$. Since D_2 is dense, there is some $g_2 \in D_2$, $g_2 \leq g_1$. And so on. Then $G^* = \{x : \exists g \in G(x \geq g)\}$ is the desired filter.

The Rasiowa–Sikorski lemma is of fundamental significance in the theory of forcing. Martin's axiom, which is really a way to summarize lots of forcing arguments in one combinatorial principle, is almost a generalization of the Rasiowa–Sikorski lemma.

Martin's Axiom. If X is a ccc partial order and $\mathcal{D}$ is a family of fewer than 2^{ω} many dense subsets of X, then there is a $\mathcal{D}$-generic filter on X.

We abbreviate Martin's axiom as MA and note two variations of it: $MA + \neg CH$ is the statement "MA holds and CH fails." $MA_{\aleph_1}$ is the statement "If X is a ccc partial order and $\mathcal{D}$ is a family of at most ω_1 many dense subsets of X, then there is a $\mathcal{D}$-generic filter on X." Note that by lemma 53 CH implies MA, and $MA + \neg CH$ implies $MA_{\aleph_1}$ which, by exercise 22, implies $\neg CH$. MA is independent of $ZFC + \neg CH$, that is, there are models of $ZFC + \neg CH$ in which MA holds (Solovay, Martin) and others in which it fails.

We will prove

Theorem 54. Assume MA. Then

(a) 2^{ω} is a regular cardinal.
(b) Every base for an ultrafilter on ω has size 2^{ω}.
(c) Every dominating family in ${}^{\omega}\omega$ has size 2^{ω}.

We have stated theorem 54 as an analogue of theorem 45. Note, however, the contrast between CH and $MA + \neg CH$: Under CH every base for an ultrafilter and every dominating family has size ω_1; under $MA + \neg CH$, no base for an ultrafilter and no dominating family has size ω_1.

Theorem 55. Assume MA. Then if $\omega < \kappa < 2^\omega$, $2^\kappa = 2^\omega$. In particular, under MA + ¬CH, $2^\omega = 2^{\omega_1}$, again in contrast to CH.

Theorem 56. Assume $MA_{\aleph_1}$. Then there are no Suslin trees.

Theorem 56 is due to Solovay and Tennenbaum. Theorem 54(b) is due to Booth; the rest of theorems 54 and 55 are due to Solovay and Martin. We prove them in order of ease of proof.

Proof of Theorem 56. Suppose there is a Suslin tree. Then, by exercise 16, there is a Suslin tree T so that every element has successors of arbitrarily high height. That is, for $\alpha < \omega_1$, if $D_\alpha = \{t \in T:$ height $t \geq \alpha\}$, then each D_α is dense under $\leq^{-1}$. By hypothesis there is some filter G in the sense of $\leq^{-1}$ meeting each D_α. But then G is uncountable and is a branch, so T is not Suslin after all.

Proof of Theorem 54(b). Let $\mathscr{A}$ be a family of subsets of ω of size less than 2^ω where every finite subset of $\mathscr{A}$ has infinite intersection. We will show that $\mathscr{A}$ has a lower bound. This will prove the theorem, since any family of subsets of ω either has size less than 2^ω or has size 2^ω, and no base for a nonprincipal ultrafilter has a lower bound.

We get our lower bound by means of a partial order $\mathbb{P}$ defined as follows:

Elements of $\mathbb{P}$ are ordered pairs (a, F) where a is a finite subset of ω and F is a finite subset of $\mathscr{A}$. The order is $(a, F) \leq (a', F')$ iff $a \supset a'$, $F \supset F'$, and $a - a' \subset \bigcap F'$. That is, every new element of a is a member of each set in F'.

The reader should check that this really is a partial order. We will check that it is ccc: Note that each (a, F) is compatible with each (a, F'), since no new elements are added to a. Since there are only countably many finite subsets of ω, we have $\mathbb{P}$ a countable union of pairwise compatible sets, hence $\mathbb{P}$ cannot have an uncountable antichain.

Now we need a small collection of dense sets. (In the presence of MA, "small" always means "size less than 2^ω.") For each $A \in \mathscr{A}$ let $D_A = \{(a, F): A \in F\}$. Each D_A is dense: Given (a, F) note that $(a, F) \geq (a, F \cup \{A\}) \in D_A$.

We also need, for each finite n, $D_n = \{(a, F): |a| \geq n\}$. Each D_n is dense: Given (a, F) let $b \subset \bigcap F$ where $|b| = n$. Then $(a, F) \geq (a \cup b, F) \in D_n$.

Since $|\mathscr{A}| < 2^\omega$, by MA we have some filter G meeting each D_A and each D_n. Let $A^* = \bigcup \{a:$ some $(a, F) \in G\}$. We must show that $A^* - A$ is finite, for all $A \in \mathscr{A}$, and that A^* is infinite.

Pick $(a, F) \in G \cap D_A$. Then for each $n \in A^* - a$, $n \in \bigcap F$, so, in particular, $n \in A$. Thus $A^* - A \subset a$, which is finite.

As for the cardinality of A^*, since G meets each D_n, $|A^*| \geq n$ for all finite n. So A^* is infinite.

Proof of Theorem 54(c). The proof is similar to that for theorem 54(b): We show that every small family of functions is dominated via an appropriate partial order, hence a dominating family cannot be small.

So suppose F is a family of functions in $^\omega \omega$, $|F| < 2^\omega$. We define the partial order $\mathbb{P}$ as follows:

Elements of $\mathbb{P}$ are pairs (s, E) where s is a function from some n into ω and E is a finite subset of F. We say $(s, E) \leq (s', E')$ iff s extends s', $E \supset E'$ and, for all $f \in E'$ and all $n \in \mathrm{dom}\ s - \mathrm{dom}\ s'$, $s(n) > f(n)$.

$\mathbb{P}$ is ccc since any two (s, E), (s, E') are compatible; hence, $\mathbb{P}$ is the union of countably many pairwise compatible subsets, so has no uncountable antichains.

For each finite n we let $D_n = \{(s, E): n \in \mathrm{dom}\ s\}$. For each $f \in F$ we let $D_f = \{(s, E): f \in E\}$. The reader can check that each D_n and each D_f is dense. By hypothesis we have fewer than 2^ω of them, so by MA there is some filter G which meets every D_n and every D_f. Let h be defined by $h(n) = k$ iff there is some $(s, E) \in G$ with $s(n) = k$. Since G is a filter, h is well-defined. Since G meets each D_n, $\mathrm{dom}\ h = \omega$. Let $(s, F) \in D_f \cap G$. Then if $n \notin \mathrm{dom}\ s$, $h(n) > f(n)$. So h dominates each f in F.

Theorem 54(a) follows from theorem 55: If CH holds, 2^ω is regular; if CH fails, since $2^{\mathrm{cf}(2^\omega)} > 2^\omega$, by theorem 55 $\mathrm{cf}(2^\omega)$ cannot be less than 2^ω.

So the last thing to do is prove theorem 55. First some more definitions.

Definition 57. A collection of infinite subsets of a countable set X is said to be an almost disjoint family on X iff any two distinct elements of the collection have finite intersection.

For example, if T is the binary tree of height ω and B is the set of branches of T, then B is almost disjoint, since any two distinct branches have at most finitely many elements of T in common.

Lemma 58. There is an almost disjoint family on ω of size 2^ω.

Proof. Let f be a 1–1 onto function from the binary tree T of height ω to ω. Let B be the set of branches of T. $|B| = 2^\omega$ since every function in $^\omega 2$ defines a unique branch. Define $g(b) = \{n: f^{-1}(n) \in b\}$. Then $\{g(b): b \in B\}$ is the desired family.

Definition 59. Let $\mathcal{A}$ be an almost disjoint family on ω, and let $\mathcal{C} \subset \mathcal{A}$, $\mathcal{C} \neq \mathcal{A}$. The partial order $\mathbb{P}_{\mathcal{A},\mathcal{C}}$ consists of all pairs (a, E) where a is a finite subset of ω and E is a finite subset of $\mathcal{C}$. The order on $\mathbb{P}_{\mathcal{A},\mathcal{C}}$ is defined as follows: $(a, E) \leq (a', E')$ iff $a \supset a'$, $E \supset E'$, and $(a - a') \cap (\bigcup E') = \emptyset$, i.e., any new element of a is not in any set in E'.

Note that any two (a, E), (a, E') are compatible, hence $\mathbb{P}_{\mathcal{A},\mathcal{C}}$ is ccc and, for each $A \in \mathcal{C}$, $D_A = \{(a, E): A \in E\}$ is dense. Since $\mathcal{A} - \mathcal{C}$ is nonempty, no finite union from $\mathcal{C}$ equals ω or covers any element of $\mathcal{A} - \mathcal{C}$, and each $D_n = \{(a, E): |a| \geq n\}$ is dense for finite n. For $A \in \mathcal{A} - \mathcal{C}$, $D_{A,n} = \{(a, E): |a \cap A| \geq n\}$ is dense.

Lemma 60 (Solovay's Lemma). Assume MA. Let $\mathcal{A}$, $\mathcal{C}$ be as in definition 59. If $|\mathcal{A}| < 2^\omega$, then there is a set $A_\mathcal{C} \subset \omega$ so that, for each $A \in \mathcal{A}$, $A \cap A_\mathcal{C}$ is finite iff $A \in \mathcal{C}$.

Proof. We have fewer than 2^ω many D_n's, D_A's, and $D_{A,n}$'s, so by MA there is some filter G meeting all of them. Let $A_\mathcal{C} = \bigcup\{a: \text{ some } (a, E) \in G\}$. By the D_A's, $A_\mathcal{C} \cap A$ is finite for all $A \in \mathcal{C}$; by the $D_{A,n}$'s, $A_\mathcal{C} \cap A$ is infinite for all $A \in \mathcal{A} - \mathcal{C}$.

The partial order $\mathbb{P}_{\mathcal{A},\mathcal{C}}$ is known as almost disjoint forcing, and has many applications apart from the one we are about to give.

Proof of Theorem 55. Suppose $\omega < \kappa < 2^\omega$. Let $\mathcal{A}$ be an almost disjoint family of size κ. ($\mathcal{A}$ exists by Lemma 58.) There are exactly 2^κ many subsets $\mathcal{C}$ of $\mathcal{A}$, and for each $\mathcal{C}$ we have $A_\mathcal{C} \subset \omega$ as in lemma 60. If $\mathcal{C} \neq \mathcal{C}'$ and $A \in \mathcal{C} - \mathcal{C}'$, then $A_\mathcal{C} \cap A$ is finite and $A_{\mathcal{C}'} \cap A$ is infinite. Hence no two $A_\mathcal{C}$'s are the same. Hence $2^\omega \geq 2^\kappa$, and we are done.

Now let us show that the conclusion of theorem 47 fails under MA + ¬CH.

Theorem 61 (Baumgartner–Hajnal). Assume MA + ¬CH, and suppose $\omega \times \omega_1 = H \cup K$. Then there is an infinite set A and an uncountable set B, so either $A \times B \subset H$ or $A \times B \subset K$.

Proof. For each $\alpha < \omega_1$, let $A_\alpha = \{n: (n, \alpha) \in H\}$. Notice that if $B_\alpha = \omega - A_\alpha$, then $B_\alpha = \{n: (n, \alpha) \in K\}$. Let $\mathcal{F}$ be a nonprincipal ultrafilter on ω. For every α, either A_α or $B_\alpha \in \mathcal{F}$. Either uncountably many A_α's are in $\mathcal{F}$, or uncountably many B_α's are in $\mathcal{F}$. Without loss of generality, assume the former, and let $E = \{A_\alpha: A_\alpha \in \mathcal{F}\}$. By the proof of theorem 54(b), E has a lower bound A. Now, for $A_\alpha \in E$, let n_α be the least n, so

$A - A_\alpha \subset n$. There is an uncountable set B and a fixed n with $n_\alpha = n$ for all $\alpha \in B$. But then $(A - n) \times B \subset H$, as desired.

SECTION 7.6. STATIONARY SETS AND $\diamond$

In this final section we discuss a combinatorial principle which strengthens CH. This principle, known as $\diamond$, holds in L and was formulated by Jensen as a crystalization of his proof that Suslin trees exist in L (just as MA was formulated as a crystalization of the proof that Suslin trees need not exist). L is a rich source of powerful combinatorial principles, with fanciful ad hoc names such as $\clubsuit$ (club, a weakening of $\diamond$: $\diamond = \clubsuit + CH$), $\square$ (box), and the existence of a morass. Any of Devlin's books on constructibility will have material on all of these, while further discussion of $\diamond$ and $\square$ can be found in Jech's *Set Theory*.

Before stating $\diamond$ we need to characterize some interesting sets of ordinals.

Definition 62. Let κ be a cardinal.

(a) A subset A of κ is said to be closed iff, for all $B \subset A$, if B is not cofinal in κ then sup $B \in A$.

(b) A club subset of κ is a closed subset cofinal in κ. ("Club" is short for "closed unbounded.")

For example, all subsets of ω are closed, so every unbounded subset of ω is a club. If κ is uncountable, then $\{\alpha < \kappa : \alpha$ is a limit ordinal$\}$ is a club, since the limit ordinals are cofinal in κ and every supremum of a subset of limit ordinals is itself a limit ordinal.

Notice that if A is club in κ and $\text{cf}(\kappa) = \lambda$, then A must contain elements of every cofinality below λ: Every cofinal set has order type $\geq \lambda$, so if ρ is regular and $\rho < \lambda$, then each closed unbounded set has a subset of order type ρ whose sup has cofinality ρ.

Notice that if $\text{cf}(\kappa) > \omega$ then the intersection of two clubs is a club. (In fact something stronger is true—see exercise 26.)

Definition 63. Let κ be a cardinal. A subset A of κ is said to be stationary iff $A \cap C \neq \emptyset$ for all club subsets C of κ.

For example, if κ has uncountable cofinality, then by the note above $\{\alpha < \kappa : \text{cf}(\alpha) = \omega\}$ is a stationary set, and every club is stationary. If κ has cofinality ω, then S is a stationary subset of κ iff $\kappa - S$ is bounded below κ.

The division of sets into stationary and nonstationary sets is an important one. To give an idea of the power of stationary sets, we state two theorems about them.

Theorem 64 (Fodor). Let κ be a regular uncountable cardinal. If f is a function on a stationary set $S \subset \kappa$ and $f(\alpha) < \alpha$ for all nonzero $\alpha \in S$ (such a function is called a regressive function), then there is a stationary $S' \subset S$ with f constant on S'.

Theorem 65 (Solovay). Every regular uncountable cardinal κ is the union of κ many disjoint stationary subsets of κ.

We will prove Fodor's theorem. The reader is referred to any standard advanced set-theory text for the proof of Solovay's theorem.

Proof of Theorem 64. Suppose f is regressive on S, where S is stationary in some regular uncountable κ. If the theorem fails, then for all $\alpha < \kappa$ there is a club C_α where f misses α on $C_\alpha \cap S$; i.e., if $\beta \in C_\alpha \cap S$, then $f(\beta) \neq \alpha$.

Let $C = \{\beta : \forall \alpha < \beta (\beta \in C_\alpha)\}$. Let us show that C contains a club. Define $g : \kappa \to \kappa$ by $g(\gamma) = \inf \bigcap_{\alpha < \gamma} C_\alpha$. Without loss of generality, g is increasing.

Claim. The function g is continuous.

Proof. Suppose $\gamma = \sup\{\gamma_i : i \in \lambda\}$ where $\lambda < \kappa$. Let $\delta = \sup\{g(\gamma_i) : i \in \lambda\}$. Since each $g(\gamma_i) \in \bigcap_{\alpha < \gamma_i} C_\alpha$, $\delta \in \bigcap_{\alpha < \gamma_i} C_\alpha$ for all i. Hence $\delta \in \bigcap_{\alpha < \gamma} C_\alpha$. Since each $g(\gamma_i) \notin \bigcap_{\alpha < \gamma_{i+1}} C_\alpha$, $\delta = \inf \bigcap_{\alpha < \gamma} C_\alpha$. So $\delta = g(\gamma)$.

Returning to the proof of theorem 64, an adaptation of the proof of theorem 28 in Chapter 5 shows that every increasing continuous function on κ has a club of fixed points. So let D be a club of fixed points for g. Then D is contained in C. Since C contains a club and S is stationary, there is some nonzero $\beta \in C \cap S$. Since $\beta \in C_\alpha$ for all $\alpha < \beta$, $f(\beta) \geq \beta$, a contradiction.

As another example of the importance of stationary sets, consider this more general version of Silver's singular cardinals theorem (theorem 46, chapter 5).

Theorem 66 (Silver). Suppose $\aleph_\alpha$ is a singular cardinal of uncountable cofinality and, for some fixed $\gamma < \mathrm{cf}(\alpha)$, there is set S stationary in α so that, for all $\beta \in S$, $2^{\aleph_\beta} = \aleph_{\beta+\gamma}$. Then $2^{\aleph_\alpha} = \aleph_{\alpha+\gamma}$.

Stationary sets and their generalizations permeate modern set theory—for example, they are crucial in forcing and large cardinal theory—but we will give only one further example of their use, in the principle $\diamond$, which holds in L.

$\diamond$. There is a sequence $\{A_\alpha : \alpha < \omega_1\}$ so that $A_\alpha \subset \alpha$ and, for each set $A \subset \omega_1$, $\{\alpha : A \cap \alpha = A_\alpha\}$ is stationary.

We call the sequence $\{A_\alpha : \alpha < \omega_1\}$ a $\diamond$-sequence and say that it captures each $A \subset \omega_1$ on a stationary set.

Notice that under $\diamond$ each countable subset of ω_1 equals stationarily many A_α's.

Notice that $\diamond$ implies CH: Since each subset of ω equals stationarily many A_α's, we can have at most ω_1 of them.

We use $\diamond$ to get a Suslin tree.

Theorem 67. Assume $\diamond$. Then there is a Suslin tree.

Proof. Let $\{A_\alpha : \alpha < \omega_1\}$ be a $\diamond$-sequence. We will use this sequence to construct a Suslin tree T whose elements will be countable ordinals. Let $\Lambda = \{\delta_\alpha : \omega \le \alpha < \omega_1\}$ enumerate all countable limit ordinals. Note that Λ is a club. Before giving the construction we need one more definition: An antichain in a partial order is maximal iff every element in the partial order is compatible with some element of the antichain. By the axiom of choice, every antichain in a partial order extends to a maximal antichain.

We will construct T so

(*)   each $T_\alpha \subset \delta_\alpha$ and T_α is a splitting tree.

Start out by letting T_ω be any countable splitting tree of height ω whose elements are the finite ordinals. Now suppose, for infinite α, we have T_α. We check A_α. Is it a maximal antichain in the tree T_α? If not, we do whatever we like to extend T_α to $T_{\alpha+1}$ so (*) is not violated. If A_α is a maximal antichain in T_α, then, for each element β of A_α we let b_β be a branch of T_α with $\beta \in b_\beta$. We have chosen a countably infinite set of such b_β's and $\delta_{\beta+1} - \delta_\beta$ is countably infinite, so we can extend T_α to $T_{\alpha+1}$ so each new element of $T_{\alpha+1}$ is an upper bound for some b_β. Since every element added to T from then on has height $> \alpha$ and hence is comparable to some element in $T(\alpha)$, A_α will be a maximal antichain for T.

Since T is a splitting tree, it suffices to show that it has no uncountable antichains. So suppose A is an uncountable antichain of T. A extends to a maximal antichain, so since we are trying to show that A does not exist we might as well assume that A is itself maximal. Consider $C = \{\alpha: A \cap T_\alpha$ is a maximal antichain in $T_\alpha\}$. We show that this set is a club in ω_1. Clearly C is closed.

Suppose $\alpha < \omega_1$. For each element t in $T_{\alpha+1}$ we pick some $a_t \in A$ with t, a_t comparable (note that t may equal a_t). Let $\gamma_0 = \sup(\{\alpha + 1\} \cup \{\beta:$ for some t in $T_{\alpha+1}$, $a_t \in T(\beta)\})$. Now for each t in T_{γ_0+1}, pick some $a_t \in A$ with t, a_t comparable and let $\gamma_1 = \sup(\{\gamma_0 + 1\} \cup \{\beta:$ for some t in T_{γ_0+1}, $a_t \in T(\beta)\})$. And so on. Finally, if $\gamma = \sup\{\gamma_n: n < \omega\}$, then $\gamma \in C$ and $\gamma > \alpha$. Thus C is unbounded. (This sort of argument is called a Lowenheim–Skolem closure argument.)

Since $\{A_\alpha: \alpha < \omega_1\}$ is a $\diamondsuit$-sequence, there is some $\delta_\alpha \in C \cap \Lambda$ with $A \cap \delta_\alpha = A_{\delta_\alpha}$. A_{δ_α} is maximal in T_{δ_α} since $\delta_\alpha \in C$. Thus, by construction of T, $A = A_{\delta_\alpha}$ contradicting that A is uncountable.

This is a classic use of $\diamondsuit$: You kill off every possible counterexample by killing off some reflection in the $\diamondsuit$-sequence.

The general technique of reflection—in order to check something at a high level you look at its reflections down below—permeates not only infinite combinatorics on small cardinals (e.g., $\diamondsuit$) but also is a key element of certain large cardinal principles (e.g., supercompact cardinals) and forcing techniques (e.g., proper forcing). One of the major themes of set theory in the 1980s has been the importance of stationary sets and reflection, so it is fitting that these techniques end this book.

EXERCISES FOR CHAPTER 7

1. (a) Show that the statements in example 2(a) are true.
   (b) Show that in the partition of example 2(b) cubics are not homogeneous.

2. A weak antichain in a partial order is a set of pairwise incomparable elements. Use Ramsey's theorem to show that every infinite partial order has either an infinite chain or an infinite weak antichain.

3. Use Ramsey's theorem to show that every infinite set of integers has an infinite subset X so that either (1) if $n < m$ are elements of X then n divides m or (2) if $n < m$ are elements of X then n does not divide m.

4. Use Ramsey's theorem to show that every infinite set of natural numbers has an infinite subset X so that either every three distinct elements in X sum up to a prime or no three distinct elements in X sum up to a prime.

5. Use Ramsey's theorem to show that if A is a finite set of primes then every infinite set of natural numbers has an infinite subset X so that either there is some fixed p in A where every four distinct elements of X sum up to a multiple of p or no four distinct elements of A sum up to a multiple of any element of A.

6. Use Ramsey's theorem to show that every infinite linearly ordered set has either an infinite increasing sequence or an infinite decreasing sequence.

7. Use the Erdős–Rado Theorem to show that if a partial order has size $>2^\omega$ then it has either an uncountable chain or an uncountable weak antichain (see exercise 2 for definition of weak antichain).

8. Consider the set of increasing well-ordered sequences of elements of $\mathbb{R}$, ordered by end-extension ($g \geq f$ iff f is an initial segment of g). Show that this is a tree. What is its height? What is the size of each level?

   Note: A well-ordered sequence is a function from some ordinal α into $\mathbb{R}$. The ordinal α need not equal ω.

9. We say a tree T is a tree of subsets of some set X iff (1) every element of T is a subset of X, (2) $a \leq_T b$ iff $a \supset b$, (3) if a, b are incompatible in T then $a \cap b = \emptyset$. Suppose T is a tree of subsets of X. Show

   (a) T has no collection $\{a_n : n < \omega\}$ where each $a_n \subset a_{n+1}$.

   (b) $|\text{height } T| \leq |X|$.

   (c) Each level of T has cardinality at most $|X|$.

   (d) It is possible to have height $T > |X|$.

10. Using König's lemma, show that if the human race is to survive forever then some woman must have a female descendant in each subsequent generation.

11. Let T be a tree whose elements are ordinals where $\alpha <_T \beta$ implies $\alpha > \beta$ and each level of T is finite. Show that T is finite.

12. Let T be a tree whose elements are sets where $a <_T b$ implies $b \in a$ and each level of T is finite. Show that T is finite.

13. Show that every countable linear order extends to a countable dense linear order with no endpoints.

14. (a) Show that an uncountable subset of a Suslin tree is Suslin.

    (b) Show that an uncountable subset of a Suslin line need not be Suslin.

    (c) Show that if there is a Suslin line then there is one whose every uncountable subset is Suslin.

15. Show that a tree of height ω_1 which is the union of countably many antichains is not Suslin.

16. Say that a Suslin tree is nearly normal iff every node has uncountably many sucessors.

  (a) Show that if T is a nearly normal Suslin tree then every element of T has at least two incompatible successors.

  (b) Show that if there is a Suslin tree then there is a nearly normal Suslin tree. (Method 1: Attach a copy of the original tree to every node which has only countably many successors. Method 2: Throw out all nodes with only countably many successors.)

17. An Aronszajn line is any linear order extending that of an Aronszajn tree. Show

  (a) If an Aronszajn line extends a special Aronszajn tree, then the line has uncountably many pairwise disjoint intervals.

  (b) No Aronszajn line embeds in an order-preserving fashion into $\mathbb{R}$.

18. Show directly, without using theorem 39, that every measurable cardinal is regular.

19. Let κ be measurable and let j, M be as in theorem 40. Show

  (a) The statement "$j(\kappa)$ is measurable" holds in M.

  (b) If κ is the smallest measurable cardinal in V, then "κ is not measurable" holds in M.

20. Let j be as in theorem 40, and show that if $x \subset y$ then $j(x) \subset j(y)$.

21. Assume CH. Show that there is a tree of size ω_1 and height ω_1 with at least ω_2 many uncountable branches. (Such a tree is called a Canadian tree.) (*Hint:* it is a standard example). *Note:* The consistency of "ZFC + there are no Canadian trees" is equiconsistent with the consistency of "ZFC + there is an inaccessible cardinal."

22. (a) Let T be a tree. Show that T has no filter generic for all D_C, where D_C is defined as in the paragraph after definition 52.

  (b) Show that $M A_{\aleph_1}$ implies $\neg$CH.

23. Use the Rasiowa–Sikorski lemma to reprove theorem 46.

24. An almost disjoint family $\mathscr{A}$ is maximal iff every infinite subset of ω has infinite intersection with some element of $\mathscr{A}$.

  (a) Show that under CH every maximal almost disjoint family on ω has cardinality ω_1.

  (b) Show that under MA every maximal almost disjoint family on ω has cardinality 2^ω.

25. A scale is a dominating family in ${}^\omega\omega$ well-ordered by $\leq^*$. Show that

  (a) Under CH there is a scale of order-type ω_1.

  (b) Under MA there is a scale of order-type 2^ω.

26. Show that if $cf(\kappa) = \lambda > \omega$ then the intersection of fewer than λ many club subsets of κ is a club subset of κ.

27. Let κ be regular, $f: \kappa \to \kappa$, f continuous and $|\text{range } f| = \kappa$. Show that $\{\alpha: f(\alpha) = \alpha\}$ contains a club.

28. Show that if κ has countable cofinality then $S \subset \kappa$ is a stationary subset of κ iff $\kappa - S$ is bounded below κ.

29. Let $\{A_\alpha: \alpha < \omega\}$ be a $\diamondsuit$-sequence. Show that for no countable set A is $\{\alpha: A = A_\alpha\}$ a club.

BIBLIOGRAPHY

Necessarily, a book at this level leaves a lot out. The reader interested in learning more might find a selected bibliography helpful, so here it is. The books are divided into categories. Some acquaintance with the model-theory portions of the logic books is essential for understanding nearly all of the set-theory books mentioned.

MATHEMATICAL LOGIC

Enderton, H., *A Mathematical Introduction to Logic*, Academic Press, New York, 1972.

Shoenfield, J., *Mathematical Logic*, Addison-Wesley, New York, 1967.

SET THEORY (GENERAL)

van Dalen, D., Doets, H. C., and de Swart, H., *Sets: Naive, Axiomatic and Applied*, Pergamon, London, 1978.

Henle, J., *An Outline of Set Theory*, Springer-Verlag, New York, 1986.

Jech, T., *Set Theory*, Academic Press, New York, 1978.

FORCING

Burgess, J., "Forcing," in *Handbook of Mathematical Logic*, J. Barwise, ed., North-Holland, Amsterdam, 1977.

Kunen, K., *Set Theory: An Introduction to Forcing and Independence Proofs*, Studies in Logic and the Foundations to Mathematics, Vol. 102, North-Holland, Amsterdam, 1980.

Jech, T., *Lectures in Set Theory*, Lecture Notes in Mathematics, Vol. 217, Springer-Verlag, Berlin and New York, 1971.

COMBINATORICS

Erdős, P., Máté, A., Hajnal, A., Rado, R., *Combinatorial Set Theory: Partition Relations for Cardinals*, Studies in Logic and the Foundations of Mathematics, Vol. 106, North-Holland, Amsterdam, 1984.

Williams, N., *Combinatorial Set Theory*, Studies in Logic and the Foundations of Mathematics, Vol. 91, North-Holland, Amsterdam, 1977.

L AND RELATED MODELS

Devlin, K., *Constructibility*, Springer-Verlag, Berlin, 1984.

Dodd, A. J., *The Core Model*, London Mathematical Society Lecture Notes Series, Vol. 61, Cambridge University Press, Cambridge, 1982.

LARGE CARDINALS

Drake, F. R., *Set Theory*, North-Holland, Amsterdam, 1974.

Kanamori, A., Reinhardt, W., Solovay, R., "Strong axioms of infinity and elementary embedding," *Annals of Math. Logic*, Vol. 13 (1978), 73–116.

Kanamori, A., Magidor, M., *The Evolution of Large Cardinal Axioms in Set Theory*, Lecture Notes in Mathematics, Vol. 669, Springer-Verlag, Berlin, 1978, pp. 99–275.

HISTORY

Dauben, J. W., *Georg Cantor: His Mathematics and Philosophy of the Infinite*, Harvard University Press, Cambridge, Massachusetts, 1979.

van Heijenoort, J., ed., *From Frege to Gödel: A Source Book in Mathematical Logic, 1879–1931*, Harvard University Press, Cambridge, Massachusetts, 1967.

Moore, Gregory H., *Zermelo's Axiom of Choice: Its Origins, Development and Influence*, Springer-Verlag, Berlin, 1982.

Directory of Symbols

Symbol	Definition	Page
$\mathbb{N}$	set of natural numbers	2
$\mathbb{Z}$	set of integers	—
$\mathbb{Q}$	set of rationals	—
$\mathbb{R}$	set of reals	—
$\in$	is an element of	25
$\subset$	subset	3, 25
$\cup$	union	25, 34
$\cap$	intersection	25, 34
$\mathscr{P}$	power set	41
ω	$\mathbb{N}$	41
$\aleph$	(aleph)	80
$\beth$	(bet)	80
$\rightarrow$	(arrow notation for partition calculus)	113
R^{-1}	inverse relation	33
V_α	αth initial segment of V	55

INDEX